Control of Fluid-Containing Rotating Rigid Bodies

Communications in Cybernetics, Systems Science and Engineering

ISSN: 2164-9693

Book Series Editor:

Jeffrey Yi-Lin Forrest
International Institute for General Systems Studies, Grove City, USA
Slippery Rock University, Slippery Rock, USA

Volume 2

Control of Fluid-Containing Rotating Rigid Bodies

Anatoly A. Gurchenkov, Mikhail V. Nosov & Vladimir I. Tsurkov

Department of Complex Systems, Computing Center of the Russian Academy of Sciences, Moscow, Russia

CRC Press is an imprint of the
Taylor & Francis Group, an **informa** business
A BALKEMA BOOK

CRC Press
Taylor & Francis Group
6000 Broken Sound Parkway NW, Suite 300
Boca Raton, FL 33487-2742

First issued in paperback 2019

Typeset by MPS Limited, Chennai, India

ISBN-13: 978-1-138-00021-6 (hbk)
ISBN-13: 978-0-367-38017-5 (pbk)

Library of Congress Cataloging-in-Publication Data

Gurchenkov, Anatoly A.
Control of fluid-containing rotating rigid bodies / Anatoly A. Gurchenkov, Mikhail V. Nosov & Vladimir I. Tsurkov.
pages cm
Includes bibliographical references and index.
ISBN 978-1-138-00021-6 (hardback)
1. Rotational motion (Rigid dynamics) 2. Motion control devices. 3. Tanks—Automatic control.
4. Sloshing (Hydrodynamics) I. Title.
QA862.G9G87 2013
531'.34—dc23

2013007040

Visit the Taylor & Francis Web site at
http://www.taylorandfrancis.com

and the CRC Press Web site at
http://www.crcpress.com

Table of contents

Editorial board

Notations

x	scalar variable x;
$\boldsymbol{y}$	vector variable $\boldsymbol{y}$;
J, B, L	tensors and operators (matrices);
$(\boldsymbol{x}, \boldsymbol{y})$	scalar product of the vectors $\boldsymbol{x}$ and $\boldsymbol{y}$;
$\boldsymbol{x} \times \boldsymbol{y}$	vector product of the vectors $\boldsymbol{x}$ and $\boldsymbol{y}$;
$\dot{x}$	time derivative of x;
$\ddot{x}$	second time derivative of x;
f', f''	spatial derivatives of f;
u^*	complex conjugate to u;
V_0	initial value of the variable V in the Cauchy problem;
$\Vert\boldsymbol{x}\Vert$	Euclidean norm of the vector $\boldsymbol{x}$;
$i = 1, \ldots, n$	i ranges from 1 to n;
$u \in U$	u belongs to the set U;
$u \notin U$	u does not belong to the set U;
$U \subset H$	U is a subset of the set H;
$\mathbb{R}^n$	Euclidean n-space;
$[a, b]$	closed interval;
$[a, b)$	half-open interval;
(a, b)	open interval.

About the authors

Anatoly A. Gurchenkov, Doctor of Science (in physics and mathematics), Leading Researcher at the Department of Complex Systems, Computer Center, Russian Academy of Sciences and Professor at the Russian State Technological University (MATI). A well-known specialist in stability and control of rotating dynamic systems with a fluid-filled cavity. Author of more than 100 papers published by the Russian Academy of Sciences, "Fizmatlit" and other publishing houses, including four monographs; author of two patents.

Mikhail V. Nosov, Candidate of Science (in physics and mathematics), Senior Researcher at the Department of Complex Systems, Computer Center, Russian Academy of Sciences. Author of more than 25 papers published by the Russian Academy of Sciences and other publishing houses. Teaches courses in programming, computer science, and theory of relational databases at the Russian State Technological University (MATI).

Vladimir I. Tsurkov, Doctor of Science (in physics and mathematics), Head of the Department of Complex Systems Computer Center, Russian Academy of Sciences, and Professor at the Moscow Institute of Physics and Technology. Member of the Institute for Management Sciences and Operations Research (USA), Associate Member of the Russian Academy of Natural Sciences, Member of the Editorial Board of the *Journal of Computer and Systems Sciences International.* One of the leading experts in aggregation-based decomposition methods and a well-known specialist in analysis of numerical methods related to operation research models, in large-scale hierarchical optimization and control, and in catastrophe theory. Author of more than 180 papers published by the Russian Academy of Sciences, "Fizmatlit" and other publishing houses (in particular, Kluwer), including six monographs.

Introduction

Dedicated to the blessed memory
of Viktor Mikhailovich Rogovoi

The emergence of control theory is associated with the names of Maxwell and Vyshnegradskii. They formulated the purpose of control—ensuring stability—and proposed methods for selecting parameters of control systems. Since then, over almost 100 years, basic problems of control theory turned out to be related, in some respect, to stability problems. In the 1920–1930s, in the theory of regulation (which was the name of control theory at that time), ideas and methods of Lyapunov stability became widespread; in turn, control theory continuously stimulated the interest of researchers in stability problems.

In the 1930–1940s, objects of study were steady regimes, the stability of aircraft motion, stabilization of turbine turns, and so on. In the 1950s, engineering demands had required analyzing unsteady processes.

The short paper [243] by Okhotsimskii, which contains a solution of the first problem of optimal control theory, opened a new page in the application of the theory.

The general formalism of optimal control theory, which has received the name of the maximum principle, was created by L. S. Pontryagin and his students V. G. Boltyanskii, R. V. Gamkrelidze, and E. F. Mishchenko. The maximum principle has made it possible to reduce the optimal control problem to a special boundary value problem.

In the late 1950s–early 1960s, L. S. Pontryagin, R. Bellman, R. Kalman, N. N. Krasovskii, and their collaborators laid the foundations of modern mathematical control theory [18, 20, 26, 156, 180, 182, 183, 254], some of whose postulates had previously been stated at the engineering level. A special role in the formation and development of this theory was played by Pontryagin's maximum principle, Bellman's method of dynamic programming, and the theory of linear control systems.

Some of the formed direction of optimal control theory are as follows. The references given below do not pretend to be comprehensive; a more complete bibliography can be found therein. Thus, these directions are control under conditions of uncertainty [192]; control in conflict situations [184]; control of systems with distributed parameters [312]; hierarchical control systems [167, 295, 312]; construction of sets of attainability [44]; control and stabilization with respect to a part of variables [7, 173, 174, 267, 320]; control of artificial intelligence systems [316]; and control in the presence of uncontrolled noise [179, 183, 321]. Also one may point out the works [8, 13, 18–21, 26, 67, 149, 150, 155–158, 167, 173–175, 179–184, 192, 224, 254, 312, 316, 317].

At present, the theory and solution methods of linear control problems have been developed fairly extensively. The solution of nonlinear problems causes significant difficulties. In this respect we mention the method based on the application of Pontryagin's maximum principle to optimal control problems with "movable ends," methods of game theory in the case of control in the presence of uncontrolled noise [184], the asymptotic method [5, 8], the method of oriented manifolds [173, 175], and the method of nonlinear transformations of variables combined with a special choice of the structure of controls [319, 320].

The applications of these methods cover, in particular, a number of problems of mechanics, such as the problem of controlling the motion of an aircraft [5, 7, 8], the nonlinear version of the speed optimization problem for the rigid collision of two material points [6], the problem of damping the rotation of an asymmetric rigid body by means of one balance wheel [173], and the problem of controlling the orientation of an asymmetric rigid body [321].

This study is devoted to an interesting and difficult mechanical problem concerning the rotational motion of a body with a cavity filled with a fluid. On a twisted body, longitudinal moments of forces act along the principal axis which cause precession motion. In the settings of an ideal and a viscous fluid in the cases of partial and complete filling, we were able to obtain integral dependences of longitudinal angle velocities on these moments, which play the role of controls. In turn, these dependences determine the stability of the motions under consideration. To derive these basic relations, we solve intermediate problems on the motion of a fluid in a body, as well as of the body itself. Then, we state a large class of optimal control problems with functionals involving angular velocity. Applying a number of transformations, we obtain systems to which the Hamilton–Pontryagin formalism and Bellman's optimality principle apply.

The problem of the motion of a rigid body with a cavity completely filled with a fluid began to attract attention of scientists as early as the middle of the nineteenth century. Stokes was, apparently, the first to be interested in this important mechanical problem (1842–1847; see [305]); it was also considered by Helmholtz [139] and Lamb [197], who investigated a number of special cases, and by Neumann [241] (1883), who was interested in this problem in relation to the study of the motion of a rigid body in a fluid. In their research, the problem of oscillations of a bounded volume of fluid appeared as a problem of standing wave theory.

The first detailed study of the dynamics of a rigid body containing a cavity completely filled with a homogeneous incompressible fluid in a general setting was performed by Zhukovskii [332] in 1885. It was shown that the irrotational flow of the fluid in the cavity is determined by the motion of the body, and the latter coincides with the motion of the same body in which the fluid is replaced by an equivalent rigid body. Determining the motion of the fluid in the cavity requires solving certain stationary boundary value problems depending only on the shape of the cavity. Solutions of these problems (Zhukovskii potentials) make it possible to find the components of the associated mass tensor for the cavity under consideration.

The motion of a body with a cavity containing an ideal fluid in potential motion turns out to be equivalent to the motion of a rigid body whose inertia tensor is composed of the inertia tensor of the initial rigid body and the associated mass tensor for the cavity under consideration.

Thus, the problem of describing the motion of a fluid-containing body decomposes into two parts. The first part of the problem, which depends only on the geometry of the cavity, reduces to solving boundary value problems and to calculating the associated mass tensor. The second part of the problem is the usual problem of the motion of a rigid body, which reduces to solving a system of ordinary differential equations.

An analysis of the problem revealed a whole series of difficulties, which rendered any attempts to apply analytical methods impractical. For this reason, after the first success, the theory of fluid oscillations almost ceased to develop, all the more so because, for a long time, this theory had no direct engineering applications.

Further progress in this theory was related to the seiche problem. The researchers' attention was attracted by surprising phenomena called seiches. In large lakes water experiences periodic motion similar to tides in the ocean. But these phenomena could not be explained from the point of ebbs and flows, because seiche periods in different lakes are different. Only in the twentieth century did a relationship between the seiche phenomena and standing wave theory become clear. The development of effective methods for calculating seiche periods inside complex basins, such as lakes and shallow seas, had become possible thanks to various simplifying assumptions (such as those of shallow water theory). Simultaneously with the introduction of these simplifications, the development of various methods of numerical analysis began.

It should be mentioned that problems of continuum mechanics and hydrodynamics have always served as a stimulus for the development of new directions in mathematics and mathematical physics. An illustration of this statement is the flow of new ideas in the theory of nonlinear differential equations and the establishment of striking relations between, seemingly, different branches of mathematics, which had entailed the study of the Korteweg–de Vries equation for waves in shallow water.

It turned out that these problems, which are very instructive theoretically, are very important from the point of view of practice too. Thus, the problems of the theory of fluid oscillations attract the attention of hydraulic engineers and constructors of dock buildings. Similar problems arise in relation to the study of the seismic stability of various reservoirs for storing fluid and in the theory of motion of ships, submarines, and aircraft.

The interest in this area noticeably increased with the development of rocket and space technologies. The large amounts of liquid fuels carried by rockets, satellites, and spacecraft may substantially affect the motion of these aircrafts in certain cases.

The study of the dynamics of wave motions of various fluids is also related to problems of geophysics, oceanology, atmosphere physics, and environmental protection. For example, the problems of motion of bodies with fluid-containing cavities have found application in the study of the dynamics of spacecraft, which uniformly rotate on their orbits about some axis for the purpose of stabilization, uniform heating by solar rays, the creation of artificial gravity, and so on. These problems also arise in designing rapidly rotating rotors, spinners, and gyroscopes, which have fluid-filled cavities. Finally, the behavior of a fluid under conditions of zero or low gravity affects the behavior of space ships etc.

Simultaneously with the problem of the motion of a body with a fluid-containing cavity, the problem of the stability of such motion arose. Thus, in Kelvin's experiments [163], it was established that the rotation of a top is stable if the cavity is compressed in the direction of the axis of rotation and unstable if the top has stretched shape. The theoretical study of this problem was performed by Greenhill [81], Hough [141], Poincaré [253], and other scientists. These scientists considered the motion of a rigid body with an ellipsoidal cavity filled with an ideal fluid. The fluid experiences motions of a special kind (homogeneous vortex motions).

Hough [141] studied the characteristic equation for small oscillations of a rigid body with a fluid near uniform rotation in the case of an ellipsoidal cavity containing an ideal fluid experiencing homogeneous vortex motion.

Sobolev [299] considered the motion of a heavy symmetric top with a cavity containing an ideal fluid. The equations of motion were linearized near the uniform rotation of the top. Sobolev determined some general properties of motion, in particular, stability conditions. In [299] two special cases of a cavity were considered, an ellipsoid of revolution and a circular cylinder. A similar problem was considered by different methods by Ishlinskii and Temchenko in [148]. An experimental study of this problem was presented in [210]. Stewartson [303] studied the stability of a heavy top with a cylindrical cavity containing a fluid with free surface.

The study of the equations of motion for a rotating fluid showed (see Poincaré's work [253]) that these equations have a number of features distinguishing them from the usual equations of mathematical physics. Various mathematical questions related to the equations of a rotating fluid were considered by Aleksandryan [10], Krein [185], and other authors.

A more complicated case is where the fluid is subject to the action of surface tension forces. These forces substantially affect the equilibrium and the motion of the fluid in the case where the mass force is small, which occurs under conditions close to the absence gravity. Thus, the dynamics of a fluid subject to the action of surface tension forces is of applied interest in space engineering. In [233] and [38, 43] problems of equilibrium and motion for a fluid in a vessel in the presence of surface tension forces were considered. In particular, oscillations of an ideal fluid subject to surface tension forces were considered in [233] and oscillations of a viscous fluid in [170]. In [41] the dynamics of a rigid body with a cavity filled with an ideal fluid with an air bubble was studied. It was shown by using Schwarz' method that the motion of such a system can be described by ordinary differential equations. Purely mathematical questions of the theory of small oscillations are fairly simple: the problem reduces to an equation with completely continuous self-adjoint operator. However, the computational aspects of the theory have been poorly developed. Among works in the latter direction we mention those of Chetaev [45], Rumyantsev [265, 269–275], Pozharitzkii [255, 256], Kolesnikov [168], and others.

In [45] Chetaev gave a rigorous solution of the problem of the stability of the rotational motion of a projectile with a cavity completely filled with an ideal fluid experiencing irrotational motion. The problem was solved in the nonlinear setting.

Chetaev considered three cases. In the first case, the cavity has the shape of a circular cylinder whose axis coincides with the axis of rotation of the inertia ellipsoid of the projectile (without fluid). Using Zhukovskii's results, Chetaev showed that the problem of the stability of the rotational motion of such a projectile coincides with the classical problem of the stability of a usual projectile with appropriate moments of inertia. Using his previous results, Chetaev wrote an inequality ensuring the stability of the rotational motion of a fluid-filled projectile for flat trajectories.

In the second case, the cavity had the shape of a circular cylinder with one flat diaphragm. In this case, the ellipsoid of inertia of the projectile and the associated masses turned out to be triaxial, which impeded the application of results obtained previously for solid rigid projectiles. Chetaev wrote out Lagrange equations describing the motion and studied the stability of the unperturbed motion in the first approximation. For reduced moments of inertia, he wrote inequalities implying that the roots of the characteristic equation of the first approximation are purely imaginary and the unperturbed motion is stable in the first approximation.

In the third case, the cavity is a circular cylinder in which diaphragms compose a cross formed by two mutually orthogonal diametrical planes. In this case, Chetaev numerically obtained sufficient stability conditions for the flight of a projectile of the type under consideration. This work of Chetaev initiated the study of the stability of rotational motions of rigid bodies with cavities completely filled with a fluid or containing a fluid with free surface which experiences generally turbulent motion.

Rumyantsev obtained a series of results on the stability of motion of a rigid body with a cavity completely or partially filled with a fluid. He considered both the case of an ideal fluid and the case of a viscous fluid, as well as the influence of surface tension forces (see [272]). Rumyantsev's sufficient conditions agree with results obtained by Sobolev in [299]. Namely, for the stability of the rotation of a free rigid body with a fluid about its center of inertia, it is sufficient that the axis of rotation be the axis of the greatest central moment of inertia of the entire system [232]. This result supplements Zhukovskii's theorem from [332].

In the setting of the motion stability problem for a fluid-containing rigid body, which is a system with infinitely many degrees of freedom, the definition of the notion of stability is of fundamental importance.

Three main approaches to the study of stability in the nonlinear setting of problems have been proposed. If the motion of the fluid in the cavity is characterized by a finite number of variables, then the stability problem reduces to a problem of the Lyapunov stability of a system with many finite degrees of freedom. This simplest case occurs only if the cavity is completly filled with an ideal fluid and the motion of the fluid is irrotational, or if the fluid experiences homogeneous turbulent motion in an ellipsoidal cavity.

If the state of the system is described by infinitely many variables, then we must state the motion stability problem for a finite number of variables and introduce certain quantities being integral characteristics of the motion of the fluid. In this case, the method of Lyapunov functions can also be applied [232, 273].

The third approach is related to Lyapunov's ideas concerning the theory of equilibrium figures of a rotating fluid and leads to a generalization of Lagrange's and Routh's theorems. In this case, the problem of the stability of an equilibrium or the steady motion reduces to minimizing a certain functional.

In the case of a cavity completely filled with a fluid, stability is understood in the sense of Lyapunov with respect to noncyclic coordinates of the body, its generalized velocities, and the kinetic energy of the fluid. If the cavity is filled only partially, then by stability we understand the stability of generalized coordinates and velocities of the body, as well as the stability of the "equilibrium" shape of the fluid.

Many authors considered the dynamics of a rigid body with a cavity containing an ideal fluid with free surface. This problem is of great importance for applications.

In addition to the stability issues, of interest are joint oscillations of the fluid and the fluid-containing body, as well as the development of effective numerical methods for calculating the motion of such systems. This problem is largely considered in the linear setting.

The general problem of oscillations of a body with a cavity partially filled with an ideal fluid was studied by Moiseev [225–230], Okhotsimskii [244], Narimanov [240], Rabinovich [259], Krein and Moiseev [187], and other authors.

It turned out that describing small oscillations of a body with a cavity containing a heavy ideal fluid with free surface requires not only calculating the Zhukovskii potentials but also solving an eigenvalue problem. This problem, which depends only on the shape of the cavity, is a problem of the free oscillations of a fluid in a fixed vessel. By determining the Zhukovskii potentials and the free oscillations of the fluid, we can find coefficients characterizing the mutual influence of the body and the fluid in the cavity under oscillations. The motion of the whole system can be described by countably many ordinary differential equations, whose coefficients are determined in the way specified above. In this case, the problem again decomposes into two parts.

The first part of the problem, which depends on the geometry of the cavity, reduces to solving certain boundary value problems and eigenvalue problems for linear partial differential equations followed by calculating hydrodynamic coefficients. This problem can only be solved analytically for a few shapes of cavities. In the case of a cavity of complex shape, it is solved by using various numerical or approximate methods. These questions have an extensive literature; see, e.g., the collection of papers [315] and papers [231, 260].

The second part of the problem consists in studying and solving a system of ordinary differential equations. In practical problems, it is usually sufficient to take into account only a few basic forms of fluid oscillations, so that the number of equations describing the motion of the fluid is small. This problem can be solved numerically.

If the oscillations of the fluid in the vessel cannot be assumed to be small, the problem becomes nonlinear. Some nonlinear problems about the motion of a fluid with free surface inside a cavity in a rigid body were considered in [230, 239].

The third part of the problem, the directions specified above is related to experimental research. Experiments are performed under laboratory and real-life conditions. The long duration and high price of experimental modeling, the conditional character of the applicability of laboratory results to real tasks, and, sometimes, unsurmountable difficulties involved in setting up experiments distinguish mathematical methods of hydrodynamics.

These methods make it possible to reveal patterns of the phenomena being studied, perform a comprehensive analysis for prediction purposes, and theoretically determine parameters of the processes of wave propagation and their interaction with obstructions.

In [219] free oscillations of a fluid in a vessel were investigated; in particular, the influence of viscosity and surface tension on the oscillations was measured. In [220] oscillations of a body with a fluid were experimentally studied.

Problems of the dynamics of a rigid body with a cavity containing a viscous fluid are significantly more complicated than in the case of an ideal fluid. There is less literature on these problems. The problems are largely considered in the linear setting, and the questions under examination are either stability issues or special cases of motion of bodies with cavities of special shape.

Rumyantsev [266] considered the problem on the motion of a body with a cavity containing a viscous fluid at small Reynolds numbers. Some approximate solutions of problems on the motion of a viscous fluid in a rotating cavity are contained in [197, 296].

In the case of large Reynolds numbers (that is, the low viscosity of the fluid), a reliable solution method for hydrodynamic equations is the method of boundary layer [165, 198]. A mathematical substantiation of this method for some linear boundary value problems was given by Vishik and Lusternik in [318].

Moiseev studied the spectrum of the free oscillations of a heavy ideal fluid in a fixed vessel [225–234]. He considered a body with an oscillating fluid as a mechanical system with infinitely many degrees of freedom.

The hydrodynamic problem was separated from the problem of the dynamics of an equivalent body to which an infinite system of mathematical pendula is associated. The dynamical characteristics of a fluid with "frozen" free surface were calculated by Zhukovskii's method, and the oscillations of the free surface were determined by solving the problem of the free oscillations of a fluid in a fixed cavity.

Moiseev considered also the nonlinear setting of the problem of the oscillations of a heavy ideal fluid in a vessel. He applied an approach suggested by Poincaré. Namely, he sought an approximate solution of the problem in the form of an asymptotic series in a small parameter (the wave amplitude) and showed that the spectrum of free oscillations is piecewise continuous rather than discrete, and the oscillation amplitude can take any value in the disk of convergence of the series. Considering the problem of forced oscillations of a fluid in periodic field of mass forces, Moiseev had succeeded in constructing approximate asymptotic solutions degenerating into the trivial solution near and far from the resonance.

A cycle of papers by A. A. Petrov, a student of Moiseev, is devoted to oscillations of a bounded volume of fluid in a cavity of a rigid body. For cavities of various shapes, Petrov gave a variational setting of the problem about the motion of a fluid in a vessel of finite size and obtained approximate solutions of problems in variational setting [248]. In [250] the free oscillations of a fluid in a fixed vessel were calculated by variational methods. In [231, 249] approximate numerical methods for calculating the free oscillations of a fluid in a vessel of any shape and the Zhukovskii potentials for these vessels were proposed.

Moiseev suggested a version of the boundary layer method for studying small oscillations of a viscous fluid [229], which was used by Krasnoshchekov [178] to solve the problem of small plane oscillations of a pendulum with an axially symmetric cavity filled with a low-viscosity fluid. The method presented in [229] was applied to various problems on oscillations of a viscous fluid with free surface in [15, 189, 282].

Some general theorems on properties of the free oscillations of a heavy viscous fluid in a cavity of a rigid body were proved by Krein [186] by methods of functional analysis. Ievleva [144, 145] considered the motion of a rigid body with a spherical cavity filled with a viscous fluid and succeeded in expressing the solutions in terms of generalized spherical functions.

Research on the motion of a fluid in a cavity of a rotating rigid body was performed in many countries. One can point out some of it [11, 12, 32, 59, 60, 68–75, 151–153,

159–162, 186–191, 217, 218, 235, 236, 238, 241, 242, 245, 253, 264, 276, 278, 279, 296–298, 301, 302, 308, 310, 311, 315, 330, 331, 333–335]. We outline works by Greenspan and Howard [84], Greenspan [86, 87], Stewartson [303], Stewartson and Roberts [304]. The following books [4, 22–25, 27, 28, 30, 46, 48, 52–54, 63–66, 76–80, 83, 88, 140, 157, 163–166, 168, 169, 171, 185, 247, 257, 262, 263, 265, 280–282, 295, 305, 312, 314, 322, 324–329] are used here.

In these works, the motion of the bodies under consideration was assumed to be either a uniform motion [84–87] or a regular precession [304]. The method of boundary layer (two-scale decomposition) was extensively applied. In [85] the coincidence of results of computations by the boundary layer method with experimental data was mentioned. The most general results were obtained by Chernous'ko [36], who used the method of asymptotic integration.

Chernous'ko considered a large set of related problems in [33–43]. More specifically, he considered the motion of a rigid body with a cavity completely filled with a viscous incompressible fluid. For the nonlinear setting of the problem of in-plane motion of a pendulum with a cavity filled with a viscous fluid, he obtained a law of attenuation for nonlinear oscillations and rotations of a pendulum. He also investigated the stability of a spatial motion of a free rigid body with a cavity filled with a viscous fluid and studied the motion of a rigid body with a cavity completely filled with a low-friction incompressible fluid (the Reynolds number was assumed to be large). To solve the linearized Navier–Stokes equations, Chernous'ko applied the boundary layer method. He obtained general formulas determining a solution in the boundary layer in terms of the Zhukovskii potentials. In addition, he studied small oscillations of a rigid body with a cavity filled with a viscous fluid. The shape of the cavity and the number of degrees of freedom of the rigid body were assumed to be arbitrary, and the oscillations were studied in the linear setting. Both the free oscillations under the action of conservative external moments as forced oscillations were considered. The cases of a fluid with low and high viscosity were investigated in parallel. For these cases, asymptotic solutions were obtained and compared the exact solution for an infinite cylindrical cavity.

Chernous'ko also studied the motion of a rigid body with a cavity partially filled with an incompressible viscous fluid with free surface (the viscosity was assumed to be low). He considered some problems on small oscillations of a rigid body with a cavity partially filled with a low-viscosity fluid. Moreover, he obtained formulas for the eigenfrequencies and attenuation coefficients of the free oscillations of a body with a fluid and calculated the amplitudes of forced oscillations of a fluid and a fluid-containing body. All considerations were performed for cavities of arbitrary shape.

Rotational motions of a fluid-containing body close to a uniform rotation about an axis were considered. The Navier–Stokes equations were linearized near the uniform rotation. Special solutions of linearized equations for the turbulent motion of an ideal fluid were introduced. These solutions, which depend on the shape of the cavity, are similar to the Zhukovskii potentials for the case of an irrotational motion. Tensors similar to the associated mass tensor, which characterize the influence of the fluid in the cavity on the motion of the rigid body, were considered. For ellipsoidal and spherical cavities, special solutions of the turbulent motion equations were determined and the components of the tensors mentioned above were calculated. Some problems concerning the dynamics of rotational motions of a body with a fluid were also considered.

The characteristic equation for the oscillations of a free rotating rigid body with a cavity filled with an ideal or viscous fluid were composed. In the case where the mass of the fluid is small in comparison with the mass of the body, this equation was solved for any shape of the cavity. Problems on the stability of the free motion were studied, and the roots of the characteristic equation were calculated, which made it possible to judge the rate of the amplification or attenuation of oscillations.

Rvalov and Rogovoi [277] considered the linear setting of the Cauchy problem for a motion of a body with an ideal fluid perturbed with respect to a uniform rotation. They proposed an approach based on jointly solving the equations of hydrodynamics and mechanics for a cavity of any shape.

The cycle of papers [90–135] by Gurchenkov is devoted to the study of the behavior of a viscous fluid in a cavity having the shape of an arbitrary body of revolution. They obtained the spectrum of the free oscillations of a viscous fluid completely of partially filling a cavity in a rotating vessel; the Navier–Stokes equations were linearized, and the motion was assumed to be strongly turbulent. Gurchenkov considered the case of a weakly perturbed rotation; viscosity was taken into account by the boundary layer method. He obtained expressions for the eigenvalues and eigenfunctions in the problem of the oscillations of a viscous fluid in a vessel of any shape. Gurchenkov also determined the velocity field in the boundary layer and obtained expressions for generalized dissipative forces caused by viscosity. He showed the influence of viscosity on the motion of the body in comparison with the case of an ideal-fluid filling. Moments of friction forces for various shapes of cavities were obtained. On the basis of integro-differential equations, Gurchenkov investigated the stability of the steady rotation of a dynamically symmetric body with a viscous fluid. The obtained results generalize results of [57] to the case of low viscosity. A new stability criterion in the presence of viscosity was found. Gurchenkov also studied systems with partial filling and considered cavities with constructive elements of the type of radial and annular edges. He derived expressions for the rate of energy dissipation in a cavity and obtained velocity and pressures fields for an ideal fluid experiencing perturbed motion by the Fourier method. Finally, he considered the problem of filtration, in which the fluid flows out of the vessel through a hole.

This brief review of results on the dynamics of a body with a fluid-containing cavity does not pretend to be complete. Thus, we did not mention works on oscillations of a fluid in a cavity with elastic walls. We should also mention the works [1–3, 9, 11–14, 16, 17, 29, 31, 46, 47, 49–51, 61, 81, 82, 84–87, 89, 138, 142, 154, 195, 196, 200–204, 209–213, 216, 222, 223, 234, 237, 239, 240, 243, 244, 246, 248, 283–294, 299, 300, 303, 304, 306, 309, 313]. More comprehensive surveys, as well as an extensive bibliography, are contained in [231, 232, 315] and in the review papers [29, 49, 283].

Thus, for the problem of the motion of a rigid body with a cavity containing an ideal fluid, both a general theory and effective computational methods have been developed. This refers to the case of a potential motion of the fluid, which either completely fills the cavity or has free surface but performs small oscillations.

The motion of bodies with cavities containing a viscous fluid have been studied much less, although it plays an important role in problems concerning, e.g., the dynamics of space ships and other aircraft, the calculation of the motion of such vehicles with respect to their center of mass, and the stabilization and control of their motion.

For instance, of interest is the damping action of a viscous liquid in a cavity on the motion of the rigid body containing this cavity. The influence of the fluid viscosity turns out to be fairly sophisticated: it may lead to the stabilization of the motion of the rigid body or, on the contrary, to the loss of stability.

Certainly, a detailed description of the large set of physical phenomena related to the dynamics of bodies filled with a viscous fluid requires well-developed mathematical models, which usually turn out to be very complicated, nonlinear, and multiparameter, so that these phenomena can only be efficiently studied in detail by numerical methods with the aid of modern computers. This approach requires solving the boundary value problem for the Navier–Stokes equations describing the motion of the fluid in parallel with integrating the equations of motion for the rigid body. This task is very laborious can hardly be accomplished. Moreover, from the applied point view, of greatest interest are integral characteristics of the motion of the fluid in the cavity and the influence of the fluid on the dynamics of the rigid body rather than a detailed description of the motion of the fluid in the cavity. Therefore, the development of efficient methods for analyzing and calculating the motion of a body with a fluid is of interest.

The contents of the chapters are as follows.

In the first chapter, the problem of rotational motions of a solid body with a cavity containing an ideal incompressible fluid is stated and studied. The shape of the cavity is arbitrary, and on the character of the perturbed motion no constraints are imposed. Simultaneously solving the equations of hydrodynamics and mechanics is reduced to solving an eigenvalue problem depending only on the geometry of the cavity and subsequently integrating a system of ordinary differential equations whose coefficients are determined in terms of solutions of boundary value problems independent of time. The first section of the first chapter presents equations of the perturbed motion of a solid body with a cavity completely filled with an ideal incompressible fluid. The boundary and initial conditions of the problem are stated. The hydrodynamic part of the problem is reduced to an infinite system of linear differential equations for the generalized Fourier coefficients of the fluid velocity. The second section studies the stability of a uniform rotation of a body with a fluid-filled cavity on the basis of ordinary differential equations whose coefficients are expressed in terms of solutions of hydrodynamic boundary value problems depending only on the geometry of the cavity. For a freely rotating body with a fluid-filled cavity, stability domains are constructed in the plane of dimensionless parameters. In the third section, the system of equations describing the perturbed motion of a solid body with an ideal fluid is reduced, by means of the Laplace transform, to an integral relation between the angular velocity and the external moment, which is treated as a control. In the fourth section, this relation is written in an equivalent form as a linear system of sixth-order ordinary differential equations.

The obtained system of equations makes it possible to solve a large class of problems of the optimal control of the motion of fluid-filled solid bodies. The fifth section contains an example with a discontinuous control. The problem is solved by using Pontryagin's maximum principle. In the sixth section of the first chapter, the application of Bellman's dynamic programming method is demonstrated. In the concluding, seventh, section, a reduction of the integral relation to a system of fourth-order linear differential equations is proposed, which is more convenient for setting and solving optimal control problems with the use of the Hamilton–Pontryagin apparatus and investigating the attainability set of the body–fluid system.

The second chapter considers the dynamics of the rotational motion of a solid body with a cavity containing an ideal fluid and a gas at constant pressure. The first section of the second chapter considers the basic equations describing the dynamics of the wave motions of a fluid being in a state of uniform rotation. Equations (together with boundary conditions) describing the dynamics of the processes under consideration are derived. The corresponding boundary value problems are the object of study in the subsequent sections. Thus, the second section considers small oscillations of a fluid in a rotating vessel and establishes some properties of the free oscillations of a fluid. In the third and fourth sections, the problem of the oscillations of a fluid in a uniformly rotating vessel is solved in the linear approximation. For a cylindrical cavity partially filled with an ideal fluid, the hydrodynamic problem is solved by the variable separation method. The problem of the perturbed motion of a rotating solid body containing a fluid with free surface is solved in the sixth section by the Bubnov–Galerkin method. The seventh section studies the stability of the free rotation of a body with a fluid. Dependences of the constructed coefficients in the inertial couplings on the numbers l and p are constructed. It is shown that the characteristic equation in this case is a third-degree polynomial. In the eighth section, the system of equations describing the perturbed motion of a solid body containing a fluid with free surface is written in an equivalent form as an integral relation between the perturbed angular velocity of the system and the moment of external forces (the controlling moment).

The third chapter has a special place in the study. Its results are used in the last chapter. In the third chapter, initial boundary value problems describing a nonstationary flow of a viscous fluid in a half-space over a rotating flat wall are stated and solved. A fluid, together with a wall bounding it, rotates at a constant angular velocity about a direction not perpendicular to the wall. The nonstationary flow is induced by the longitudinal oscillations of the wall. The velocity field and the tangential stress vector acting from the fluid on the plate are found. The solution is given in analytic form. The first section presents an exact solution of the Navier–Stokes equations in a half-space filled with a viscous incompressible fluid over a rotating flat wall performing longitudinal oscillations with speed $u(t)$. The second section studies the longitudinal quasi-harmonic oscillations of the plate and analyzes the boundary layers adjacent to the wall. The structure of the boundary layers is investigated. In the fifth section, in studying the oscillatory solutions, the resonance case is considered (the frequency of the oscillations of the plate equals twice the frequency of rotation of the plate–fluid system). Resonance leads to a nontrivial physical effect: the amplitude of oscillating velocity field does not tend to zero at infinity but remains bounded. The sixth section presents an analytic solution of the problem described above generalized to the case of a porous plate, through which the fluid can flow. In this practically important case, the structures of the velocity field and the boundary layers are described too.

The fourth chapter of the book considers the rotational motion of a solid body with a cavity completely filled with a viscous fluid. General equations of the weakly perturbed motion of a solid body with a viscous fluid are derived. The dynamics of small-amplitude waves are described. For the initial approximation the solution of the problem of the oscillations of a body with a cavity containing an ideal fluid is taken. The velocity field of a viscous fluid in a boundary layer is determined. It is shown that the tangential component of this velocity field coincides with the velocity field of the fluid in the problem of the oscillations of a flat wall in a half-space filled with a viscous incompressible fluid in which the plate velocity is replaced by the negative velocity of the ideal fluid. This result makes it possible to use the flat wall model for taking into account the influence of viscosity on the motion of a solid body containing a viscous fluid.

The first section studies small-amplitude waves of a viscous fluid in a cavity of a rotating body. In the second section, the flat wall model is used to derive a system of integro-differential equations describing the weakly perturbed motion of a body with a viscous fluid. This system of equations is solved in the fourth section by the Laplace transform method. The sixth section studies the stability of a freely rotating body with a viscous fluid. The characteristic equation of the body–fluid system is obtained and solved by means of perturbation theory. The flat wall model is used in the seventh section to study the librational motion of a solid body with a cavity containing a viscous fluid and a gas at constant pressure. Inside the cavity, there are constructive elements having the shape of radial and circular edges. Equations of the weakly perturbed motion of such a system are derived. The eighth and ninth sections study the optimal control of the perturbed motion of a solid body with a cavity containing a viscous fluid.

Thus, in the eighth section, the integral Laplace transform is applied to reduce the system of integro-differential equations with singular kernel to an integral equation coinciding in form with the integral equation obtained in the first chapter for the rotational motion of a body with a cavity containing an ideal fluid. But the roots of the equation and its coefficients, which depend only on the geometry of the cavity, have a different meaning. The ninth section outlines optimal control problems for the case of a viscous fluid.

The book is devoted to the construction of an adequate mathematical model for internal problems of the dynamics of a solid body interacting with a fluid and problems of the optimal control of these systems. These models are based on asymptotic solutions of field problems involving boundary layers of both types (time and spatial), which is their important special feature. The mathematical models constructed in the book make it possible to describe complicated transition processes. The corresponding formalism is "tuned" to the possibility of closing the mathematical model of a control object by some control equation.

Unfortunately, because of space limitations, we could not cover the whole spectrum of problems, the more so that our main purpose was to throw a bridge from control theory to the theory of a rotating body with a fluid. In each chapter exposition is largely independent of the other chapters. Notation is reintroduced in each chapter.

The monograph represents results of studies begun in the 1970s. The initiator and inspirer of this work was Viktor Mikhailovich Rogovoi, who worked at the Central Research Institute of Machine Building at that time. The untimely decease of this talented scientist and excellent man prevented him from participating in the study of control of fluid-filled rotating bodies. The ideas of Rogovoi and the mathematical apparatus developed by him have made it possible realize a number of settings of related control problems. We dedicate our book to the blessed memory of Viktor Mikhailovich, our splendid colleague and friend.

Chapter 1

Control of a rotating rigid body with a cavity completely filled with an ideal fluid

This chapter considers a mathematical model for rotational motions of a body with a cavity containing an ideal fluid. Harmonic oscillations of a fluid in a cavity were studied in [34, 148, 299]. In [148, 299], only cavities of special shape were considered. In [34], no constraints on the shape of cavities were imposed, but it was assumed that the natural vibrations of the fluid in the cavity in a rotating body were already damped out, and only the forced motion of the fluid was considered.

We do not impose any constraints on the shape of the cavity and the character of the perturbed motion of the fluid. We reduce simultaneously solving the equations of hydrodynamics and mechanics to solving an eigenvalue problem depending only on the geometry of the cavity and subsequently integrating a system of ordinary differential equations whose coefficients are determined from time-independent solutions of boundary value problems. Such a separation of the time coordinate from the spatial coordinates makes it possible to consider any perturbed motion of the body and solve boundary value problems for cavities of any configuration.

Section 1 studies the Cauchy problem for the perturbed (relative to uniform rotation) motion of a body with a cavity filled with an ideal fluid. The problem is solved in a linear setting. Using the Galerkin method, we reduce the hydrodynamic part of the problem to an infinite system of linear differential equations for the generalized Fourier coefficients of the fluid velocity.

In Section 2, we investigate the stability of a uniform rotation of a body with a fluid-filled cavity on the basis of ordinary differential equations whose coefficients are expressed in terms of solutions of hydrodynamic boundary value problems depending only on the geometry of the cavity. For a freely rotating body with a fluid-filled cavity, we construct stability regions in the plane of dimensionless parameters.

In Section 3, we reduce the system of equations describing the perturbed motion of a solid body with an ideal fluid to an integral relation between the angular velocity and the external moment by means of the Laplace transform. In Section 4, we write this relation in an equivalent form as a linear system of sixth-order ordinary differential equations. The system of equations thus obtained makes it possible to solve a large class of optimal control problems related to the motion of solid bodies with a fluid-filled cavity.

Section 5 contains an example of an optimal control problem with a discontinuous control. This problem is solved by using Pontryagin's maximum principle.

In Section 6, the integral relation is directly used in solving optimal control problems on the basis of Bellman's principle.

Section 7 proposes a universal reduction of the integral relation to a system of fourth-order linear differential equations. The form of equations of motion considered in this section turns out to be convenient for studying the attainability set of the body–fluid dynamic system.

1 EQUATIONS OF MOTION OF A RIGID BODY WITH A CAVITY COMPLETELY FILLED WITH AN IDEAL INCOMPRESSIBLE FLUID

Consider the motion of a solid body with a cavity Q completely filled with an ideal incompressible fluid of density ρ in a field of mass forces with potential U. We write the equations of motion of the fluid in the body coordinate system $Oxyz$ as

$$\begin{cases} W_0 + \boldsymbol{\omega} \times (\boldsymbol{\omega} \times \boldsymbol{r}) + \dot{\boldsymbol{\omega}} \times \boldsymbol{r} + 2\,\boldsymbol{\omega} \times V + \dfrac{\partial V}{\partial t} + (V, \nabla)V = -\dfrac{\nabla P}{\rho} - \nabla U, \\ \operatorname{div} V = 0 \quad \text{on } Q, \quad (V, \boldsymbol{n}) = 0 \quad \text{on } S, \\ V = V_0(\boldsymbol{r}) \quad \text{at } t = 0, \end{cases} \tag{1.1}$$

where t denotes time, $\boldsymbol{r}$ is the radius-vector relative to the origin O, $V = V(\boldsymbol{r}, t)$ is the fluid velocity in the coordinate system $Oxyz$, $P = P(\boldsymbol{r}, t)$ is the fluid pressure, W_0 is the absolute acceleration of the point O, $\boldsymbol{\omega} = \boldsymbol{\omega}(t)$ is the absolute angular velocity of the solid body, $\dot{\boldsymbol{\omega}}$ is the angular acceleration of the body, S is the boundary of the domain Q, and $\boldsymbol{n}$ is the unit outer normal vector to S (see Fig. 1.1).

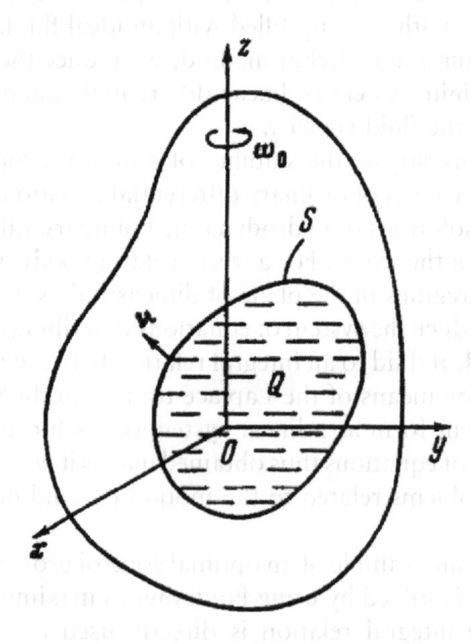

Figure 1.1 A body with a fluid.

In the body coordinate system, the equation of the moments of the entire system about the center of inertia O_1 has the form

$$\frac{dK}{dt} + \omega \times K = M_1, \quad K = J\omega + \rho \int_Q r \times V dQ, \tag{1.2}$$

where J is the inertia tensor of the body–fluid system with respect to O_1, M_1 is the resultant moment of all external forces acting on the body about the center of inertia and K is the angular momentum.

Suppose that the unperturbed motion of the body with a fluid with respect to the center of inertia is a uniform rotation of the entire system as a solid body about an axis parallel to the Oz axis at a constant angular rate ω_0.

For the unperturbed motion, we have

$$\omega = \omega_0 = \omega_0 e_z, \quad V = 0, \quad M_1 = \omega_0 \times J\omega_0.$$

Suppose that

$$\begin{aligned} &\omega(t) = \omega_0 + \boldsymbol{\Omega}(t), \quad M_1 = \omega_0 \times J\omega_0 + M, \\ &P = \rho\left[p + \frac{1}{2}(\omega \times r)^2 - U - (W_0, r)\right]. \end{aligned} \tag{1.3}$$

We assume that the quantities $\boldsymbol{\Omega}$, V, p, and M involved in the unperturbed motion are first order small.

Substituting (1.3) into equations (1.1) and (1.2) and discarding quantities of higher order of smallness, we reduce the equation of motion of the fluid to the form

$$\begin{cases} \dfrac{\partial V}{\partial t} + 2\,\omega_0 \times V + \dot{\boldsymbol{\Omega}} \times r = -\nabla p, \\ \operatorname{div} V = 0 \quad \text{on } Q, \quad (V, n) = 0 \quad \text{on } S, \\ V = V_0(r) \quad \text{at } t = 0. \end{cases} \tag{1.4}$$

The equations of motion for the body with a fluid-filled cavity take the form

$$J\dot{\boldsymbol{\Omega}} + \boldsymbol{\Omega} \times J\omega_0 + \omega_0 \times J\boldsymbol{\Omega} + \rho \int_Q r \times \dot{V} dQ + \rho \int_Q \omega_0 \times (r \times V) dQ = M. \tag{1.5}$$

The dot denotes time differentiation in the system $Oxyz$.

Equations (1.4) and (1.5), together with the usual equations of motion for the center of inertia, the kinematic relations, and the initial conditions, completely describe the dynamics of the body with a fluid.

Consider the hydrodynamic problem. Let us introduce the linear transformation $L(\sigma)$ [57] defined by

$$Lb = b + \sigma^2 e_z(e_z, b) + \sigma\, b \times e_z$$

and scalar functions $\varphi(x, y, z)$ being solutions of the boundary eigenvalue problem

$$\begin{cases} \Delta\varphi + \sigma^2 \dfrac{\partial^2\varphi}{\partial z^2} = 0 \quad \text{on } Q, \\ (L\nabla\varphi, \boldsymbol{n}) = 0 \quad \text{on } S. \end{cases} \tag{1.6}$$

According to [299], problem (1.6) has countably many eigenfunctions φ_n and eigenvalues σ_n, which are dense in the domain $\mathrm{Re}\,\sigma_n = 0$, $|\sigma_n| \geqslant 1$. Consider the complex-valued vector functions $\boldsymbol{V}_n(x_1, x_2, x_3)$ defined by

$$\boldsymbol{V}_n = -\lambda_n^{-1}\left(1+\sigma_n^2\right)^{-1} L\nabla\varphi_n, \quad \lambda_n = \frac{2\omega_0}{\sigma_n}, \tag{1.7}$$

By virtue of (1.6) and (1.7), the functions $\boldsymbol{V}_n$ satisfy the equations

$$\begin{cases} \lambda_n \boldsymbol{V}_n + 2\,\boldsymbol{\omega}_0 \times \boldsymbol{V}_n + \nabla\varphi_n = 0, \\ \operatorname{div} \boldsymbol{V}_n = 0 \quad \text{on } Q, \quad (\boldsymbol{V}_n, \boldsymbol{n}) = 0 \quad \text{on } S. \end{cases} \tag{1.8}$$

It can be shown that the $\boldsymbol{V}_n$ are orthogonal in Q in the sense that

$$\int_Q (\boldsymbol{V}_n, \boldsymbol{V}_m^*) dQ = \begin{cases} 0 & \text{if } n \neq m, \\ N_n^2 & \text{if } n = m. \end{cases} \tag{1.9}$$

Here and in what follows, $*$ denotes complex conjugation.

We take the inner product of the first equation in (1.8) and $\boldsymbol{V}_m^*$; then, we perform complex conjugation of the function $\boldsymbol{V}_m$ in (1.8) and take the inner product of the result with $\boldsymbol{V}_n$. Adding the resulting equations, we obtain

$$(\lambda_n - \lambda_m)(\boldsymbol{V}_n, \boldsymbol{V}_m^*) = -(\nabla\varphi_m^*, \boldsymbol{V}_n) - (\nabla\varphi_n, \boldsymbol{V}_m^*). \tag{1.10}$$

We take the volume integral of (1.10) and apply the continuity equation and the boundary conditions (1.8). By virtue of the Gauss–Ostrogradskii theorem, we have

$$(\lambda_n - \lambda_m) \int_Q (\boldsymbol{V}_n, \boldsymbol{V}_m^*) dQ = 0,$$

which proves the orthogonality relation (1.9).

Equations (1.4) are solved by Galerkin's method in the form of a series with unknown coefficients S_n and U_n:

$$\begin{aligned} \boldsymbol{V}(x, y, z, t) &= \sum_{n=1}^{\infty} S_n(t) \boldsymbol{V}_n(x, y, z), \\ p(x, y, z, t) &= \sum_{n=1}^{\infty} U_n(t) \varphi_n(x, y, z). \end{aligned} \tag{1.11}$$

The expression $\dot{\boldsymbol{\Omega}} \times \boldsymbol{r}$ in (1.4), which is related to the motion of the body, can be represented in the form

$$\dot{\boldsymbol{\Omega}} \times \boldsymbol{r} = \sum_{n=1}^{\infty} N_n^{-2}(\boldsymbol{a}_n^*, \dot{\boldsymbol{\Omega}})\boldsymbol{V}_n, \quad \boldsymbol{a}_n = \int_Q \boldsymbol{r} \times \boldsymbol{V}_n \, dQ; \tag{1.12}$$

here,

$$N_n^2 = \int_Q |\boldsymbol{V}_n|^2 dQ.$$

Let us substitute (1.11) and (1.12) into (1.4) and make the change $2\boldsymbol{\omega}_0 \times \boldsymbol{V}_n = -\lambda_n \boldsymbol{V}_n - \nabla\varphi_n$ in the resulting relation. The application of Galerkin's procedure to the equations for the generalized Fourier coefficients of the velocity yields

$$\begin{aligned} &N_n^2[\dot{S}_n(t) - i\lambda_n S_n(t)] + (\boldsymbol{a}_n^*, \dot{\boldsymbol{\Omega}}) = 0, \\ &S_n = S_{n0} \quad \text{at } t = 0 \quad (n = 1, 2, \dots). \end{aligned} \tag{1.13}$$

Using expansion (1.11) and the second equality in (1.12), we represent the angular momentum in the form

$$\boldsymbol{K} = J\boldsymbol{\omega}_0 + J\boldsymbol{\Omega} + \rho \sum_{n=1}^{\infty} S_n(t)\boldsymbol{a}_n. \tag{1.14}$$

The substitution of (1.14) into (1.5) reduces the equation of motion of a body with a fluid to the form

$$J\dot{\boldsymbol{\Omega}} + \boldsymbol{\Omega} \times J\boldsymbol{\omega}_0 + \boldsymbol{\omega}_0 \times J\boldsymbol{\Omega} + \rho \sum_{n=1}^{\infty} [\boldsymbol{a}_n \dot{S}_n + (\boldsymbol{\omega}_0 \times \boldsymbol{a}_n) S_n] = \boldsymbol{M}. \tag{1.15}$$

Equations (1.13) and (1.15), together with the initial conditions for $\boldsymbol{\Omega}(t)$, describe the dynamics of a body with an ideal fluid.

Thus, the problem of the dynamics of a rotating body with a cavity containing a fluid decomposes into two independent problems. The first, hydrodynamic, problem reduces to solving the boundary value problem (1.6); it depends only on the geometry of the cavity and does not depend on the motion of the body. The second, dynamical, part of the problem reduces to the Cauchy problem for a system of ordinary linear differential equations (1.13), (1.15), which can be solved by known methods analytically or numerically.

If the rotation axis of the unperturbed motion of the system is simultaneously an axis of mass and geometric symmetry of the body and the cavity, then the equations can be significantly simplified. For a dynamically symmetric body, the scalar equation of motion about the Oz axis can be separated out from the other equations, and the equation of motion with respect to the Ox and Oy axes are identical. In the cylindrical coordinate system, equations (1.13) and (1.15) take the form

$$\begin{aligned} &A\dot{g} + i(C-A)\omega_0 g + 2\rho\sum_{n=1}^{\infty} a_{xn}(\dot{S}_n + i\omega_0 S_n) = M(t), \\ &C\dot{r} = M_z, \quad N_n^2(\dot{S}_n - i\lambda_n S_n) + a_{xn}^* \dot{g} = 0 \quad (n = 1,2,\dots), \end{aligned} \tag{1.16}$$

where A and C are the principal moments of inertia of the system along the Ox and Oz axes, respectively, p, q, and r are the projections of the angular velocity $\mathbf{\Omega}$ on the axes of the system $Oxyz$, $g = p + iq$, and $M = M_x + iM_y$.

2 STABILITY OF THE STEADY ROTATION OF A SOLID BODY WITH A CAVITY CONTAINING A FLUID

In this section, we investigate the stability of the steady rotation of a body with a fluid-filled cavity on the basis of ordinary differential equations whose coefficients are determined by solutions of hydrodynamic boundary value problems independent of time. Such a separation of the time coordinate from the spatial coordinates makes it possible to consider an arbitrary perturbed motion of the body and solve boundary value problems for a cavity of arbitrary shape. The eigenvalue spectrum of the boundary value problems is densely distributed over the positive and negative halves of the real axis; we study them for the example of a cylindrical cavity.

We also construct stability regions for a freely rotating body with a fluid-filled cavity in the plane of dimensionless parameters. We arrive at a conclusion about the stability of the rotation of the body with respect to the axis corresponding to the greatest moment of inertia.

Consider again equations (1.13) and (1.15) of the preceding section:

$$\begin{aligned} &J\dot{\mathbf{\Omega}} + \mathbf{\Omega} \times J\boldsymbol{\omega}_0 + \boldsymbol{\omega}_0 \times J\mathbf{\Omega} + \rho\sum_{n=1}^{\infty}[\boldsymbol{a}_n\dot{S}_n + (\boldsymbol{\omega}_0 \times \boldsymbol{a}_n)S_n] = \mathbf{M}, \\ &N_n^2(\dot{S}_n - i\lambda_n S_n) + (\boldsymbol{a}_n^*, \dot{\mathbf{\Omega}}) = 0, \\ &S_n = S_{n0} \quad \text{at } t = 0 \quad (n = 1,2,\dots). \end{aligned} \tag{2.1}$$

Let us introduce the cylindrical coordinate system

$$x_1 = R\cos\alpha, \quad x_2 = R\sin\alpha, \quad x_3 = x.$$

We represent the functions φ_n in the form of a product as

$$\varphi_n = g_n(x,R)e^{i\alpha}. \tag{2.2}$$

For other harmonics in the circular coordinate α, the coefficients $\boldsymbol{a}_n$ vanish; i.e., the mechanical system under consideration, these harmonic are not excited [58].

The components of the vector quantities (1.12) are expressed in terms of the functions g_n as

$$a_{1n} = a_{1n}^* = -ia_{2n} = ia_{2n}^* = a_n$$
$$= -\frac{\pi\rho\chi_n}{2\omega_0}\int_G \left[R\frac{\partial g_n}{\partial x} + \frac{x}{\chi_n - 1}\left(\frac{\partial g_n}{\partial R} + \frac{g_n}{R}\right)\right]RdS,$$
$$N_n^2 = \frac{\pi\rho\chi_n^2}{2\omega_0^2}\int_G \left\{\left(\frac{\partial g_n}{\partial x}\right)^2 + \frac{\chi_n^2+1}{(\chi_n^2-1)^2}\left[\left(\frac{\partial g_n}{\partial R}\right)^2 + \frac{g_n^2}{R^2}\right] + \frac{4\chi_n}{(\chi_n^2-1)^2}\frac{\partial g_n}{\partial R}\frac{g_n}{R}\right\}R\,dS. \tag{2.3}$$

Now, we multiply the first component of the vector equation (2.1) by the imaginary unit and subtract the second component from the result. Taking into account relations (2.3), we obtain the following system of equations, which is equivalent to system (2.1):

$$A\dot{\Omega} + i(C-A)\omega_0\Omega + 2\sum_{n=1}^{\infty} a_n(\dot{S}_n - i\omega_0 S_n) = M,$$
$$N_n^2(\dot{S}_n - i\lambda_n S_n) + a_n\dot{\Omega} = 0 \quad (n = 1, 2, \ldots). \tag{2.4}$$

Here $\Omega = p - iq = \Omega_1 - i\Omega_2$ and $M = M_1 - iM_2$.

For a freely rotating body (at $M = 0$), this system has the form

$$A\dot{\Omega} + i(C-A)\omega_0\Omega + 2\sum_{n=1}^{\infty} a_n(\dot{S}_n - i\omega_0 S_n) = 0,$$
$$N_n^2(\dot{S}_n - i\lambda_n S_n) + a_n\dot{\Omega} = 0 \quad (n = 1, 2, \ldots). \tag{2.5}$$

Let us determine stability conditions for a dynamically symmetric body with a fluid. The characteristic equation of the unperturbed motion (at $M = 0$) has the form

$$A\eta + (C-A)\omega_0 - \eta(\eta - \omega_0)\sum_{n=1}^{\infty}\frac{E_n}{\eta - \lambda_n} = 0, \quad E_n = \frac{2a_n^2}{N_n^2} \tag{2.6}$$

($p = i\eta$ and $p_n = i\lambda_n$).

In the first approximation, we can replace the infinite sum by the principal term ($n = 1$). We have

$$A\eta + (C-A)\omega_0 - \eta(\eta - \omega_0)\frac{E_1}{\eta - \lambda_1} = 0. \tag{2.7}$$

Let us reduce the last relation to a quadratic equation ($\Delta = C - A$):

$$A\eta(\eta - \lambda_1) + \Delta\omega_0(\eta - \lambda_1) - \eta(\eta - \omega_0)E_1 = 0,$$
$$\eta^2(A - E_1) + \eta(\Delta\omega_0 + \omega_0 E_1 - A\lambda_1) - \Delta\omega_0\lambda_1 = 0. \tag{2.8}$$

For the stability of the steady rotation, it is necessary that the roots of this equation be real; to achieve this, we require the positivity of the discriminant:

$$D = (\Delta\omega_0 + \omega_0 E_n - A\lambda_n)^2 + 4(A - E_n)\Delta\omega_0\lambda_n \geqslant 0.$$

Equating the discriminant to zero, we obtain the following equation for the boundaries of the stability zone:

$$(\Delta\omega_0 + \omega_0 E_n - A\lambda_n)^2 + 4(A - E_n)\Delta\omega_0\lambda_n = 0,$$

or

$$\left(\Delta + E_n - A\frac{\lambda_n}{\omega_0}\right)^2 + 4(A - E_n)\Delta\frac{\lambda_n}{\omega_0} = 0 \tag{2.9}$$

Reducing this equation to a quadratic equation for Δ and making the change $\lambda_n/\omega_0 = b_n$, we obtain

$$\Delta^2 + \Delta(2Ab_n + 2E_n - 4E_n b_n) + (Ab_n - E_n)^2 = 0$$
$$D = 4(E_n + b_n A - 2E_n b_n)^2 - 4(Ab_n - E_n)^2 = 16E_n(1 - b_n)b_n(A - E_n).$$

Expressions for the roots have the form

$$\begin{aligned}
\Delta_{1,2} &= -(Ab_n + E_n - 2E_n b_n) \pm 2\sqrt{E_n(1 - b_n)b_n(A - E_n)} \\
\Delta_{1,2} &= -E_n(1 - b_n) - b_n(A - E_n) \pm 2\sqrt{E_n(1 - b_n)} \\
&= -\left(\sqrt{E_n(1 - b_n)} \mp \sqrt{b_n(A - E_n)}\right)^2.
\end{aligned} \tag{2.10}$$

Note that both roots in (2.10) are real (because $b_n \leqslant 1$, and in the limit case (2.6) implies that $A' = A - \sum_{n=1}^{\infty} E_n \geqslant \geqslant 0$ as $\omega_0 \to 0$; this is the moment of inertia of an equivalent solid body) and negative.

The region instability is located between the curves determined by the positive and the negative value of the radical (2.10). Therefore, for $\Delta = C - A \geqslant 0$, the motion is always stable. Let us show that, for $\Delta < 0$, the motion is always unstable.

We return to the quadratic equation (2.9) for the discriminant. Let us solve it with respect to $\lambda_n/\omega_0 = b_n$:

$$\begin{aligned}
&A^2 b_n^2 + b_n(2\Delta A - 2E_n A - 4E_n\Delta) + \Delta^2 + E_n^2 + 2\Delta E_n = 0, \\
&D = 4(\Delta A - E_n A - 2E_n\Delta)^2 - 4A^2(\Delta^2 + E_n^2 + 2\Delta E_n) \\
&\quad = 16\Delta E_n(\Delta + A)(E_n - A) = 16\Delta E_n C(E_n - A) \geqslant 0.
\end{aligned} \tag{2.11}$$

The roots are expressed as

$$b_n^{1,2} = \frac{E_n C + \Delta(E_n - A) \pm 2\sqrt{\Delta E_n C(E_n - A)}}{A^2} = \frac{\left(\sqrt{E_n C} \pm \sqrt{\Delta(E_n - A)}\right)^2}{A^2}. \tag{2.12}$$

Both roots in (2.12) are real and positive (because $E_n \geqslant A$ and $\Delta < 0$). Since the branches of the parabola associated with the quadratic equation (2.11) are directed upward, the instability zone corresponds to the interval (b_n^1, b_n^2) between the roots, which corresponds to the negative value of the discriminant of equation (2.8). Note that $b_n^{1,2} \to -\Delta/A = 1 - C/A < 1$ as $E_n \to 0$, and since the natural frequencies b_n have dense spectrum in the domain $|b_n| \leqslant 1$ [299], it follows that we can always choose a real value b_n in the interval (b_n^1, b_n^2) corresponding to an unstable rotation of the body about an axis corresponding to the least moment of inertia.

As an example, consider (following [57]) the cylindrical cavity with unit radius and height $2h$. The boundary conditions in (1.6) take the form

$$\frac{\partial g_n}{\partial x} = 0 \quad \text{for } x = \pm h, \qquad \frac{\partial g_n}{\partial R} + \chi_n \frac{g_n}{R} = 0 \quad \text{for } R = 1.$$

Separating variables, we obtain the solution of the boundary value problem (1.6) in the form

$$g_n = g_{lp}(x, R) = \sin(k_l x)\frac{J_1(\xi_{lp} R)}{J_1(\xi_{lp})},$$

$$k_l = \frac{\pi(2l+1)}{2h}, \quad \xi_{lp} = k_l\sqrt{\chi_{lp}^2 - 1} \quad (l = 0, 1, \ldots, \; p = 1, 2, \ldots).$$

The harmonics symmetric with respect to the coordinate x are not excited. The index n represents all possible combinations of the index numbers of the longitudinal and transverse harmonics l and p. The value ξ_{lp} is the pth root of the equation

$$\xi J_0(\xi) - \left[1 \pm \sqrt{\left(\frac{\xi}{k_l}\right)^2 + 1}\,\right] J_1(\xi) = 0. \tag{2.13}$$

The eigenvalues χ_{lp} are contained in a neighborhood of the quantity

$$[h^2 p^2/(2l+1)^2 + 1]^{1/2}$$

and densely fills both the positive and the negative part $|\chi| > 1$ of the real axis.

The inertial coupling coefficients E_{lp} are calculated by using formulas (2.3); these are

$$E_{lp} = \frac{256 h^3(\chi_{lp} + 1)}{\pi^3(2l+1)^4 \chi_{lp}(\chi_{lp} - 1)[\chi_{lp}(k_l^2 + 1) - 1]}. \tag{2.14}$$

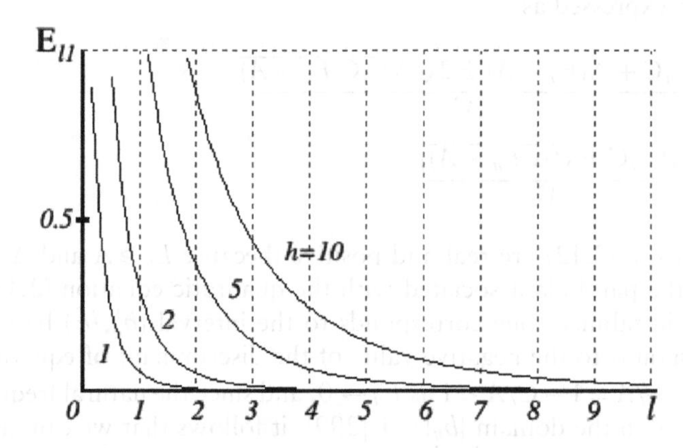

Figure 1.2 The coefficient E_{ll}.

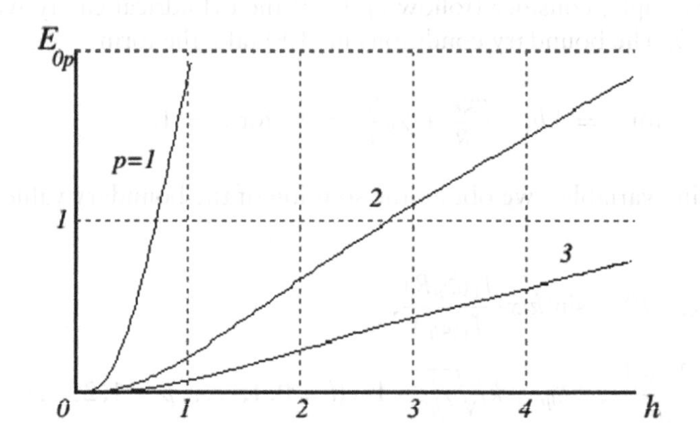

Figure 1.3 The coefficient E_{0p}.

It follows from the form of coefficients (2.14) that the series in the characteristic equation (2.6) converges, which allows us to consider approximations with finitely many terms left in the series. Figures 1.2 and 1.3 show the dependence of the inertial coupling coefficient E_{lp} on the numbers l and p, respectively, for a cylinder with $h = 1$. In most cases, it suffices to take into account only the first value $l = 0$, because the coefficient E_{lp} is inversely proportional to the fourth power of l, so that it is less by two orders for $l = 1$ than for $l = 0$. The convergence with respect to the index p is slower, to be more precise, inversely proportional to p squared. In the first approximation, the main effect appears already in the single term of the series corresponding to $l = 0$ and $p = 1$.

To reveal the influence of higher tones on the stability of the rotation of a body with a cylindrical cavity, we designed an algorithm for constructing stability regions on a computer.

The algorithm is based on the application of the Sturm method, which gives a criterion for all roots of equation (2.6) with finitely many terms in the series to be real.

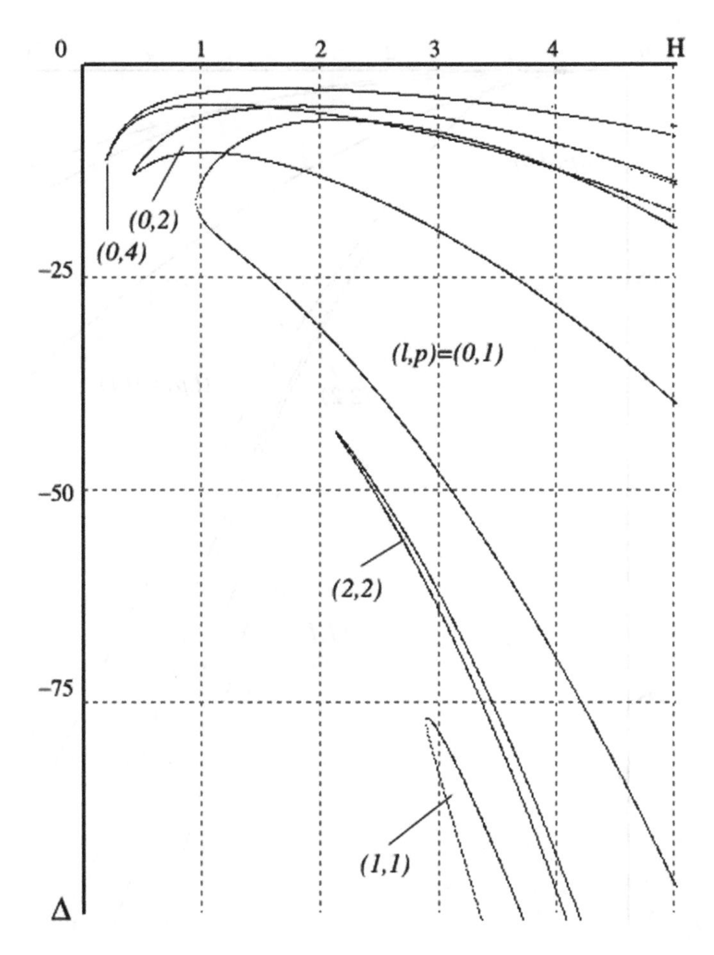

Figure 1.4 The stability region for $A^0 = 10$ and $L = 1$.

Let us introduce the dimensionless quantities $\mathbf{\Delta} = (C - A)/\rho R^5$, $h = H/R$, $2H$ (the height of the cylindrical cavity), R (the radius of its base), A^0 (the moment of inertia of the body without fluid along the transverse axis Ox reduced by ρR^5), and L (the distance from the mass center of the body with a fluid to the mass center of the fluid reduced by R).

Figures 1.4 and 1.5 show the stability regions in the dimensionless parameters $\mathbf{\Delta}$ and h with account for the following tones, being principal in the intervals of variation of $\mathbf{\Delta}$ and h under consideration: (1) $l = 0$, $p = 1 \div 4$ and $l = 1$, $p = 1, 2$ for $\chi_{lp} < 0$; (2) $l = 0$, $p = 1 \div 6$ and $l = 1$, $p = 1 \div 3$ for $\chi_{lp} > 0$. The stability regions in Figs. 1.4 and 1.5 are constructed for the parameter values $A^0 = 10$, $L = 1$ and $A^0 = 0$, $L = 0$, respectively.

It follows from these computations that, in constructing the stability region, the addition of the tones corresponding to $\chi_{lp} < 2$ only insignificantly affects this region, while taking into account tones with $\chi_{lp} > 2$ results in the emergence of new branches

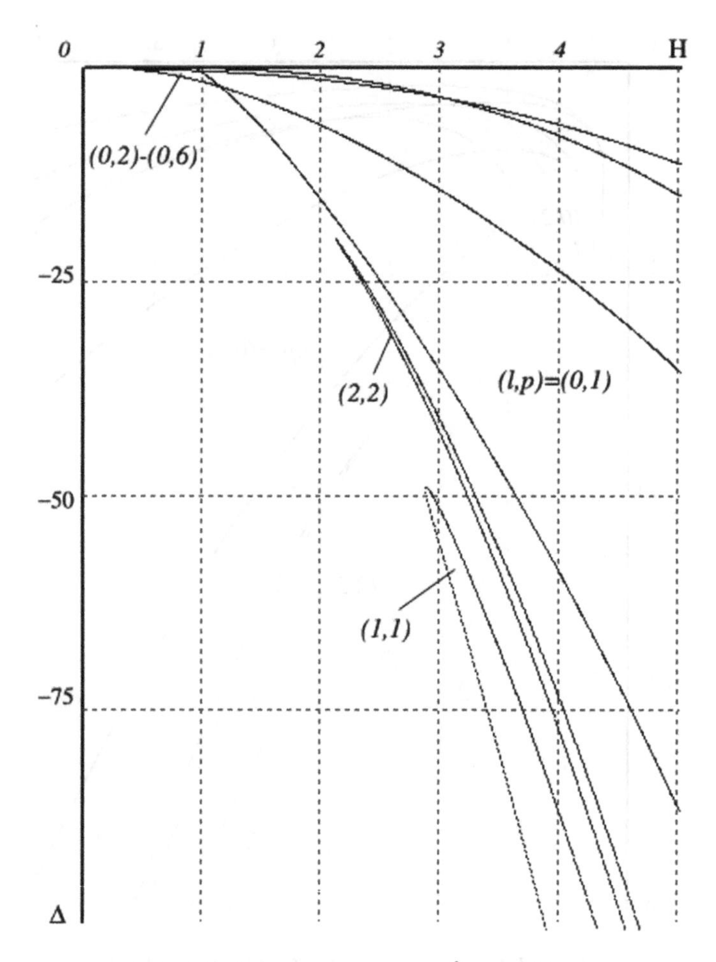

Figure 1.5 The stability region for $A^0 = 0$ and $L = 0$.

in the instability zone, which correspond to certain numbers l and p. Increasing the number p shifts the branches to the domain of smaller absolute values of $\mathbf{\Delta}$, and the branches corresponding to large l lower down to the plane of the variables $\mathbf{\Delta}$ and h.

Simultaneously increasing the numbers l and p leads to the merging of the branches corresponding to these numbers with branches corresponding to smaller l and p. For example, at $A^0 = 0$, the branches $(l = 1, p = 2)$ and $(l = 2, p = 3)$ merge with the main instability zone $(l = 0, p = 1)$. The h coordinates of the origins of branches are determined from the condition $\chi_{lp}(h) = 2$.

It should be mentioned that the increment in the oscillations of the rotation axis under the loss of stability, which is determined by the imaginary part of the root of the characteristic equation (2.6), is proportional to the value of the inertial coupling coefficient E_{lp}. Therefore, in the practical calculation of the stability region of the steady rotation of a solid body with a fluid, the number of oscillations tones of the fluid to be taken into account must be determined by comparing the hydrodynamic moments determined by the quantities E_{lp} with the value of the perturbing moment acting on the body in reality.

In the next section we consider the dependence of the angular velocity on the moment of external forces.

3 THE DEPENDENCE OF THE ANGULAR VELOCITY OF THE PERTURBED MOTION ON THE MOMENT OF EXTERNAL FORCES

In this section, we reduce, by means of the Laplace transform, the system (2.4) of differential equations describing the perturbed motion of a solid body with an ideal fluid to an integral equation for the angular velocity, which makes it possible to use the Hamilton–Pontryagin formalism for setting and solving optimal control problems.

Let us introduce the Laplace transform

$$\hat{u}(p) = \int_0^\infty e^{-pt}u(t)dt.$$

This transform reduces the system of equations (2.4) to the form

$$\begin{cases} (Ap + i(C-A)\omega_0)\hat{\Omega} + 2\sum_{n=1}^{\infty} a_n(p - i\omega_0)\hat{S}_n = \hat{M}(p), \\ N_n^2(p - i\lambda_n)\hat{S}_n + a_n p\hat{\Omega} = 0. \end{cases} \tag{3.1}$$

Here and in what follows, the hat denotes the images of the corresponding function-originals. Eliminating $\hat{S}_n$ from system (3.1), we obtain

$$(Ap + i(C-A)\omega_0)\hat{\Omega} - 2\sum_{n=1}^{\infty} \frac{a_n^2(p - i\omega_0)p\hat{\Omega}}{N_n^2(p - i\lambda_n)} = \hat{M}(p).$$

Finally, we obtain the following expression for $\Omega(p)$ as a function of the control moment:

$$\hat{\Omega}(p) = \frac{\hat{M}(p)}{Ap + i(C-A)\omega_0 - 2p(p - i\omega_0)\sum_{n=1}^{\infty} \frac{a_n^2}{N_n^2(p - i\lambda_n)}}. \tag{3.2}$$

In what follows, we consider only one summand (with $n = 1$) in the infinite sum in (3.2). We have

$$\hat{\Omega}(p) = \frac{\hat{M}(p)(p - p_1)}{(Ap + i(C-A)\omega_0)(p - p_1) - p(p - i\omega_0)E_1}, \tag{3.3}$$

where $E_1 = 2a_1^2/N_1^2$.

Consider the denominator in (3.3) separately. Its roots are

$$p^2(A - E_1) + p(i(C-A)\omega_0 - Ap_1 + i\omega_0 E_1) - i(C-A)\omega_0 p_1 = 0.$$

We set $p = i\eta$ and $\delta = C - A$. Taking into account the relation $p_1 = i\lambda_1$, we obtain

$$\eta^2(A - E_1) + \eta(\delta\omega_0 - A\lambda_1 + \omega_0 E_1) - \delta\omega_0\lambda_1 = 0. \tag{3.4}$$

The coefficients of the quadratic equation (3.4) have the form

$$a = A - E_1, \quad b = \delta\omega_0 - A\lambda_1 + \omega_0 E_1, \quad c = -\delta\omega_0\lambda_1,$$

and the discriminant of (3.4) has the form

$$D = (\delta\omega_0 - A\lambda_1 + \omega_0 E_1)^2 + 4(A - E_1)\delta\omega_0\lambda_1.$$

Accordingly, the (complex) roots of the denominator are

$$p^{(1,2)} = \frac{-(\delta\omega_0 - A\lambda_1 + \omega_0 E_1) \pm \sqrt{(\delta\omega_0 - A\lambda_1 + \omega_0 E_1)^2 + 4(A - E_1)\delta\omega_0\lambda_1}}{2(A - E_1)} i. \tag{3.5}$$

We return to the expression (3.3) for the angular velocity. We have

$$\hat{\Omega}(p) = \frac{\hat{M}(p)(p - p_1)}{(A - E_1)(p - p^{(1)})(p - p^{(2)})}, \tag{3.6}$$

where $p^{(1)}$ and $p^{(2)}$ are determined by (3.5).

To find the inverse Laplace transform, use the convolution formula

$$(f * g)(t) = \int_0^t f(\tau)g(t - \tau)d\tau.$$

In the case under consideration, we have

$$F(p) = \hat{M}(p) \quad \text{and} \quad G(p) = \frac{p - p_1}{(A - E_1)(p - p^{(1)})(p - p^{(2)})}.$$

The original of the functions in the form of the fraction $G(p) = A^n(p)/B^m(p)$, $m > n$, is determined by the formula

$$g(t) = \sum_{k=1}^{m} \frac{A^n(p^{(k)})}{B'^m(p^{(k)})} e^{p^{(k)}t},$$

where the $p^{(k)}$ are the zeros of $B^m(p)$. Returning to (3.6), we obtain

$$g(t) = \sum_{k=1}^{2} \frac{p^{(k)} - p_1}{2(A - E_1)p^{(k)} + i(\delta\omega_0 - A\lambda_1 + \omega_0 E_1)} e^{p^{(k)}t},$$

or

$$g(t) = Xe^{p^{(1)}t} + Ye^{p^{(2)}t},$$

where X and Y are known real numbers; taking into account (3.5), we obtain the following expressions for these numbers:

$$\begin{aligned}
X &= \frac{p^{(1)} - p_1}{2(A - E_1)p^{(1)} + i(\delta\omega_0 - A\lambda_1 + \omega_0 E_1)} \\
&= \frac{\eta^{(1)} - \lambda_1}{2(A - E_1)\eta^{(1)} + \delta\omega_0 - A\lambda_1 + \omega_0 E_1} \\
&= \frac{\eta^{(1)} - \lambda_1}{\sqrt{(\delta\omega_0 - A\lambda_1 + \omega_0 E_1)^2 + 4(A - E_1)\delta\omega_0\lambda_1}}, \\
Y &= \frac{p^{(2)} - p_1}{2(A - E_1)p^{(2)} + i(\delta\omega_0 - A\lambda_1 + \omega_0 E_1)} \\
&= \frac{\eta^{(2)} - \lambda_1}{2(A - E_1)\eta^{(2)} + \delta\omega_0 - A\lambda_1 + \omega_0 E_1} \\
&= -\frac{\eta^{(2)} - \lambda_1}{\sqrt{(\delta\omega_0 - A\lambda_1 + \omega_0 E_1)^2 + 4(A - E_1)\delta\omega_0\lambda_1}}.
\end{aligned}$$

Applying the convolution formula, we can write the final expression for $\Omega(t)$ as a function of the external moment $M(t)$:

$$\Omega(t) = \int_0^t M(\tau)(Xe^{p^{(1)}(t-\tau)} + Ye^{p^{(2)}(t-\tau)})d\tau. \tag{3.7}$$

4 AN EQUIVALENT SYSTEM OF EQUATIONS CONVENIENT FOR THE CONSIDERATION OF OPTIMAL CONTROL PROBLEMS

In this section we obtain a system of differential equations equivalent to relation (3.7) but more convenient for setting and solving optimal control problems by means of the Hamilton–Pontryagin apparatus.

Consider expression (3.7). We set $\Omega = \Omega_x - i\Omega_y$ and $M = M_x - iM_y$; the quantities $p^{(1)}$ and $p^{(2)}$ are defined in (3.5). For the components Ω_x and Ω_y, we can write the following expressions with real coefficients:

$$\begin{aligned}
\Omega_x(t) &= \int_0^t M_x(\tau)[X\cos\eta^{(1)}(t-\tau) + Y\cos\eta^{(2)}(t-\tau)]d\tau \\
&\quad + \int_0^t M_y(\tau)[X\sin\eta^{(1)}(t-\tau) + Y\sin\eta^{(2)}(t-\tau)]d\tau,
\end{aligned} \tag{4.1}$$

$$\Omega_y(t) = \int_0^t -M_x(\tau)[X \sin \eta^{(1)}(t-\tau) + Y \sin \eta^{(2)}(t-\tau)]d\tau$$
$$+ \int_0^t M_y(\tau)[X \cos \eta^{(1)}(t-\tau) + Y \cos \eta^{(2)}(t-\tau)]d\tau. \tag{4.2}$$

Let us introduce the notation

$$A(t) = \int_0^t [M_x(\tau)X \cos \eta^{(1)}(t-\tau) + M_y(\tau)X \sin \eta^{(1)}(t-\tau)]d\tau,$$

$$B(t) = \int_0^t [M_x(\tau)Y \cos \eta^{(2)}(t-\tau) + M_y(\tau)Y \sin \eta^{(2)}(t-\tau)]d\tau,$$

$$C(t) = \int_0^t [-M_x(\tau)X \sin \eta^{(1)}(t-\tau) + M_y(\tau)X \cos \eta^{(1)}(t-\tau)]d\tau,$$

$$D(t) = \int_0^t [-M_x(\tau)Y \sin \eta^{(2)}(t-\tau) + M_y(\tau)Y \cos \eta^{(2)}(t-\tau)]d\tau.$$

We have $\Omega_x(t) = A(t) + B(t)$ and $\Omega_y(t) = C(t) + D(t)$. Expression (3.7) is equivalent to the system

$$\dot{x}(t) = \begin{cases} \dot{\Omega}_x(t) = M_x(t)(X+Y) + \eta^{(1)}C(t) + \eta^{(2)}D(t), \\ \dot{\Omega}_y(t) = M_y(t)(X+Y) - \eta^{(1)}A(t) - \eta^{(2)}B(t), \\ \dot{A}(t) = M_x(t)X + \eta^{(1)}C(t), \\ \dot{B}(t) = M_x(t)Y + \eta^{(2)}D(t), \\ \dot{C}(t) = M_y(t)X - \eta^{(1)}A(t), \\ \dot{D}(t) = M_y(t)Y - \eta^{(2)}B(t); \end{cases}$$
$$x(0) = \begin{cases} \Omega_x(0) = 0, \\ \Omega_y(0) = 0, \\ A(0) = 0, \\ B(0) = 0, \\ C(0) = 0, \\ D(0) = 0; \end{cases} \tag{4.3}$$

We set $(n = 6, m = 2)$

$$A_{n\times n} = \begin{pmatrix} 0 & 0 & 0 & 0 & \eta^{(1)} & \eta^{(2)} \\ 0 & 0 & -\eta^{(1)} & -\eta^{(2)} & 0 & 0 \\ 0 & 0 & 0 & 0 & \eta^{(1)} & 0 \\ 0 & 0 & 0 & 0 & 0 & \eta^{(2)} \\ 0 & 0 & -\eta^{(1)} & 0 & 0 & 0 \\ 0 & 0 & 0 & -\eta^{(2)} & 0 & 0 \end{pmatrix},$$

$$B_{n\times m} = \begin{pmatrix} X+Y & 0 \\ 0 & X+Y \\ X & 0 \\ Y & 0 \\ 0 & X \\ 0 & Y \end{pmatrix}, \quad x_{0,n\times 1} = \begin{pmatrix} 0 \\ 0 \\ 0 \\ 0 \\ 0 \\ 0 \end{pmatrix}. \tag{4.4}$$

Taking into account (4.4), we rewrite system (4.3) as

$$\begin{cases} \dot{x}(t) = Ax(t) + BM(t), \\ x(0) = x_0. \end{cases} \tag{4.5}$$

Finally, we obtain a system of six equations.

In what follows we consider simplest optimal control models illustrating the applicability of the Hamilton–Pontryagin formalism and Bellman's optimality principle.

5 AN EXAMPLE WITH DISCONTINUOUS CONTROL

Consider the following optimal control problem for system (4.5):

$$J(M) = \Omega_x(T) - \gamma \int_0^T (M_x(t) + M_y(t))dt \to \max,$$
$$0 \leqslant M_x(t) \leqslant 1, \quad 0 \leqslant M_x(t) \leqslant 1 \quad \text{for any } t \in [0,T]. \tag{5.1}$$

The Hamilton–Pontryagin function has the form

$$H = -\gamma(M_x(t) + M_y(t)) + (Ax(t) + BM(t), \boldsymbol{\psi}(t)), \tag{5.2}$$

where $M(t) = (M_1(t), M_2(t))^T$ is the unknown control function and $\boldsymbol{\psi}(t)$ is the solution of the conjugate system

$$\begin{cases} \dot{\boldsymbol{\psi}}(t) = -H'_x = -A^T\boldsymbol{\psi}(t), \\ \psi_1(T) = 1, \quad \psi_i(T) = 0, \quad i = 1, \dots, 6. \end{cases} \tag{5.3}$$

Let us rewrite (5.2) in a different form (the subscript indicates the x and y coordinate components):

$$H = M_x(t)((B^T\boldsymbol{\psi}(t))_x - \gamma) + M_y(t)((B^T\boldsymbol{\psi}(t))_y - \gamma) + (A\boldsymbol{x}(t), \boldsymbol{\psi}(t)),$$

$$\begin{aligned} H = {} & (A\boldsymbol{x}(t), \boldsymbol{\psi}(t)) + M_x(t)((X+Y)\psi_1(t) + X\psi_3(t) + Y\psi_4(t) - \gamma) \\ & + M_y(t)((X+Y)\psi_2(t) + X\psi_5(t) + Y\psi_6(t) - \gamma). \end{aligned}$$

Taking into account the constraints on control (5.1), we see that the Hamilton–Pontryagin function attains its maximum for

$$\begin{cases} M_x(t) = 1 & \text{if } (X+Y)\psi_1(t) + X\psi_3(t) + Y\psi_4(t) - \gamma \geqslant 0, \\ M_x(t) = 0 & \text{if } (X+Y)\psi_1(t) + X\psi_3(t) + Y\psi_4(t) - \gamma < 0; \end{cases}$$
$$\begin{cases} M_y(t) = 1 & \text{if } (X+Y)\psi_2(t) + X\psi_5(t) + Y\psi_6(t) - \gamma \geqslant 0, \\ M_y(t) = 0 & \text{if } (X+Y)\psi_2(t) + X\psi_5(t) + Y\psi_6(t) - \gamma < 0. \end{cases} \tag{5.4}$$

Let us solve the conjugate system (5.3), which yields conditions for the switching points:

$$\psi_1(t) = 1, \quad \psi_2(t) = 0,$$
$$\begin{cases} \dot\psi_3(t) = \eta^{(1)}\psi_5(t), \\ \dot\psi_4(t) = \eta^{(2)}\psi_6(t), \\ \dot\psi_5(t) = -\eta^{(1)} - \eta^{(1)}\psi_3(t), \\ \dot\psi_6(t) = -\eta^{(2)} - \eta^{(2)}\psi_4(t). \end{cases}$$

Independently solving the equations for $\psi_3(t)$ and $\psi_5(t)$ and for $\psi_4(t)$ and $\psi_6(t)$, we obtain

$$\begin{cases} \psi_3(t) = \cos\eta^{(1)}(T-t) - 1, \\ \psi_5(t) = \sin\eta^{(1)}(T-t) \end{cases} \quad \text{and} \quad \begin{cases} \psi_4(t) = \cos\eta^{(2)}(T-t) - 1, \\ \psi_6(t) = \sin\eta^{(2)}(T-t). \end{cases}$$

Thus, the conditions on the switching points t_{M_x} and t_{M_y} of the optimal control components M_x and M_y are

$$X\cos\eta^{(1)}(T - t_{M_x}) + Y\cos\eta^{(2)}(T - t_{M_x}) = \gamma, \tag{5.5}$$

$$X\sin\eta^{(1)}(T - t_{M_y}) + Y\sin\eta^{(2)}(T - t_{M_y}) = \gamma. \tag{5.6}$$

Note that equations (5.5) and (5.6) may have no solutions; this happens, e.g., if $\gamma > |X| + |Y|$, which corresponds to the solution $M_x^* = M_y^* = 0$ (the plate γ is "too high"). They may also have countably many solutions, which corresponds to countably many intersection points of the periodic functions on the left-hand sides with the straight line $y = \gamma$.

6 APPLICATION OF BELLMAN'S OPTIMALITY PRINCIPLE

In this section we apply Bellman's optimality principle [18–21] to a particular situation with the use of the integral relation (3.7).

Consider the optimal control problem

$$J(M) = g(x(T)) + \gamma \int_{0}^{T} F(M(t))dt \to \min \tag{6.1}$$

subject to constraints of the form

$$M(t) \in U.$$

Since we know the dependence (3.7) of trajectories on the control, we can rewrite functional (6.1) in the form

$$J(M) = g\left(x(t_0) + \int_{t_0}^{T} K(\tau)M(\tau)d\tau \right) + \gamma \int_{0}^{T} F(M(t))dt \to \min, \tag{6.2}$$

where $K(t)$ is the matrix with elements $K_{i,j}(t)$ (see (7.1) and (7.2)) and

$$F(M(t)) = \|M(t)\|^2_{E^m}, \quad g(x(T,M)) = \|x(T) - y^0\|^2_{E^n}$$

for $i, j = 1, 2$; in the examples of functionals considered above, $t_0 = 0$ and $x(t_0) = 0$. The form of the functions F and g does not matter for the description of the method and can be assumed to be more general.

Let us apply Bellman's method [21]. Consider the scalar function (the Bellman function)

$$\begin{aligned} f(x,t) &= f(x_1, \ldots, x_n, t) \\ &= \min_{M(t)\in U} \left[g\left(x + \int_{t}^{T} K(\tau)M(\tau)d\tau \right) + \gamma \int_{t}^{T} F(M(\tau))d\tau \right]. \end{aligned} \tag{6.3}$$

We shall use the time grid with points $t_i = i\Delta$ and $t_{i+1} = t_i + i\Delta$ on the interval $[0, T]$; the set of i is determined by the value of the time interval under consideration. Then (6.3) satisfies the recursion relation

$$f(x,t_i) = \min_{M(t)\in U}\left[g\left(x+\int_{t_i}^{T}K(\tau)M(\tau)d\tau\right)+\gamma\int_{t_i}^{T}F(M(\tau))d\tau\right]$$

$$= \min_{M(t)\in U}\left[g\left(x+\int_{t_i}^{t_i+\Delta}K(\tau)M(\tau)d\tau+\int_{t_i+\Delta}^{T}K(\tau)M(\tau)d\tau\right)+\gamma\int_{t_i}^{T}F(M(\tau))d\tau\right]$$

$$= \min_{M(t)\in U}\left[g\left(x+\int_{t_i+\Delta}^{T}K(\tau)M(\tau)d\tau+K(t_i)M(t_i)\Delta\right)\right.$$

$$\left.+\gamma\int_{t_i}^{t_i+\Delta}F(M(\tau))d\tau+\gamma\int_{t_i+\Delta}^{T}F(M(\tau))d\tau\right]$$

$$= \min_{M(t)\in U}[\gamma F(M(t_i))+f(x+K(t_i)M(t_i)\Delta,t_i+\Delta)],$$

or

$$f(x,t_i) = \min_{M(t)\in U}[\gamma F(M(t_i))+f(x+K(t_i)M(t_i)\Delta,t_i+\Delta)]. \quad (6.4)$$

The boundary condition for (6.4) at $t_i = T$ is

$$f(x,T) = g(x(T)). \quad (6.5)$$

To solve the initial problem (6.1), according to definition (6.3), we must find the value $f(0,0)$ of the Bellman function at the initial moment of time under the initial condition

$$x(t_0) = x(0) = 0. \quad (6.6)$$

Let us solve problem (6.1) with control subject to constraints of the form $|M_l(t)| \leqslant R^2$, $l = x, y$. Consider the discrete values of the components of the control Δ_M taken with some step on the entire domain of each component (so that the number of values is of order $N_M = 2R^2/\Delta_M + 1$). There are N_M^2 possible values of the control vector at each grid point t_i, where $t_i = i\Delta$ and $t_{i+1} = t_i + i\Delta$, in the interval $[0,T]$.

We introduce a grid with step size Δ_x in the trajectory space. Since there are no constraints on the phase variables, we can construct the attainability set X_{max} of the system described by relation (7.5) at each moment of time t_i; the use of this set in the iteration procedure bounds the admissible values of phase variables. Applying (3.7), we obtain the following approximation for X_{max}:

$$x^l = \int_0^T \sum_j M_j(t)K_{l,j}(t)dt \leqslant \sum_j \int_0^T |K_{l,j}(t)|dt = x_{max}.$$

Thus, for $t_i = T$, we obtain $N_x = 2x_{max}/\Delta_x + 1$ possible values of the phase variable.

At each point of the phase space grid, for each t_i, we evaluate the Bellman function (6.4), starting the computation from the boundary condition (6.5) ($t_i = T$, $t_i = T - \Delta$, $t_i = T - 2\Delta$, …, $t_i = 0$). To optimize the iteration process and avoid an exhaustive search over all possible values of the phase variable, we consider only the values satisfying the approximation relations for the attainability set of the system. In addition to the values of the Bellman function, we shall store the value of the "optimal control" at each point satisfying (6.4). To calculate the optimal control components at the ith step, we check the belonging of the point $\boldsymbol{x} + K(t_i)\boldsymbol{M}(t_i)\boldsymbol{\Delta}$ of the attainability set obtained at the $i + 1$th step and interpolate the obtained value of the calculated point by the nearest grid value.

After completing the computation ($t_i = 0$), we shall have a set of tables of optimal control values for each point of the phase space and the values of the Bellman function at each t_i. Among all values of the phase variable at $t = 0$ we choose the value $x^l(0) = 0$ corresponding to the initial condition (6.6) of the problem under consideration. Performing reverse computation over this set of tables, we reconstruct an optimal trajectory at each moment of time, after which we obtain an optimal functional value in the initial problem.

Note that, in computing (6.4), it is sufficient to store the values of the Bellman function only at the preceding time step. The solutions are obtained in the form of a impulse curve and are constant on each interval from t to $t + \boldsymbol{\Delta}$.

In the examples considered above the functions g and F were taken in the form

$$
\begin{aligned}
g(\boldsymbol{x}) &= \sum_{l=1,2} (x^l - \Omega_l^0)^2 = (\Omega_x - \Omega_1^0)^2 + (\Omega_y - \Omega_2^0)^2, \\
F(\boldsymbol{M}(t)) &= \|\boldsymbol{M}(t)\|^2 = M_x^2(t) + M_y^2(t).
\end{aligned}
$$

Figures 1.6 and 1.7 show the solution of the problem. The control was subject to the constraint $R_l = 1$, and the terminal point was $\Omega_1^0 = 1$, $\Omega_2^0 = -1$.

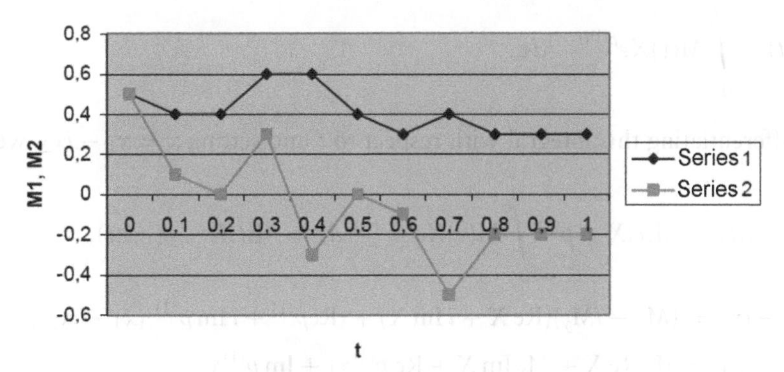

Figure 1.6 Optimal controls.

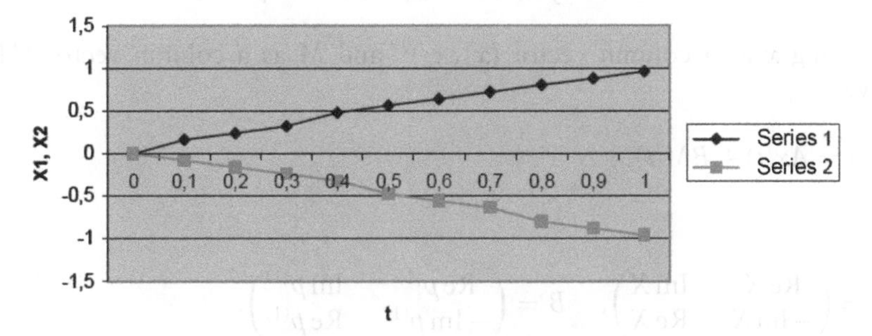

Figure 1.7 Trajectories of the system.

7 REDUCTION OF THE MAIN RELATION TO A FOURTH-ORDER SYSTEM

Below we show that relation (3.7) is equivalent to a system of four linear differential equations. Taking into account the relations $\Omega = \Omega_x - i\Omega_y$ and $M = M_x - iM_y$, we rewrite (3.7) in the form

$$\Omega_x(t) = \int_0^t M_x(\tau)K_{1,1}(t,\tau)d\tau + \int_0^t M_y(\tau)K_{1,2}(t,\tau)d\tau, \tag{7.1}$$

$$\Omega_y(t) = \int_0^t M_x(\tau)K_{2,1}(t,\tau)d\tau + \int_0^t M_y(\tau)K_{2,2}(t,\tau)d\tau, \tag{7.2}$$

where we use the obvious notation

$$\begin{aligned}
K_{1,1}(t,\tau) &= \mathrm{Re}(Xe^{p^{(1)}(t-\tau)} + Ye^{p^{(2)}(t-\tau)}),\\
K_{1,2}(t,\tau) &= \mathrm{Im}(Xe^{p^{(1)}(t-\tau)} + Ye^{p^{(2)}(t-\tau)}),\\
K_{2,1}(t,\tau) &= -K_{1,2}(t,\tau) = -\mathrm{Im}(Xe^{p^{(1)}(t-\tau)} + Ye^{p^{(2)}(t-\tau)}),\\
K_{2,2}(t,\tau) &= -K_{1,1}(t,\tau) = -\mathrm{Re}(Xe^{p^{(1)}(t-\tau)} + Ye^{p^{(2)}(t-\tau)}).
\end{aligned}$$

Consider separately the first integrand in (3.7). Let

$$x(t) = \int_0^t M(\tau)Xe^{p^{(1)}(t-\tau)}d\tau.$$

Differentiating this integral with respect to t and setting $x = x_1 - ix_2$, we obtain

$$\begin{aligned}
\dot{x}(t) &= M(t)X + p^{(1)}\int_0^t M(\tau)Xe^{p^{(1)}(t-\tau)}d\tau = M(t)X + p^{(1)}x(t),\\
\dot{x}_1 - i\dot{x}_2 &= (M_x - iM_y)(\mathrm{Re}\,X + i\,\mathrm{Im}\,X) + (\mathrm{Re}p^{(1)} + i\,\mathrm{Im}\,p^{(1)})(x_1 - ix_2),\\
\dot{x}_1 &= M_x\,\mathrm{Re}\,X + M_y\,\mathrm{Im}\,X + \mathrm{Re}\,p^{(1)}x_1 + \mathrm{Im}\,p^{(1)}x_2,\\
\dot{x}_2 &= -M_x\,\mathrm{Im}\,X + M_y\,\mathrm{Re}\,X - \mathrm{Im}\,p^{(1)}x_1 + \mathrm{Re}\,p^{(1)}x_2.
\end{aligned}$$

Treating x as a column vector $(x_1, x_2)^T$ and M as a column vector $(M_x, M_y)^T$, we have

$$\dot{x}(t) = Ax(t) + BM(t), \tag{7.3}$$

where

$$A = \begin{pmatrix} \mathrm{Re}\,X & \mathrm{Im}\,X \\ -\,\mathrm{Im}\,X & \mathrm{Re}\,X \end{pmatrix}, \quad B = \begin{pmatrix} \mathrm{Re}\,p^{(1)} & \mathrm{Im}\,p^{(1)} \\ -\,\mathrm{Im}\,p^{(1)} & \mathrm{Re}\,p^{(1)} \end{pmatrix}.$$

Similarly, for the second term in (3.7), we obtain

$$\dot{y}(t) = Cy(t) + DM(t), \tag{7.4}$$

where

$$C = \begin{pmatrix} \operatorname{Re} Y & \operatorname{Im} Y \\ -\operatorname{Im} Y & \operatorname{Re} Y \end{pmatrix}, \quad D = \begin{pmatrix} \operatorname{Re} p^{(2)} & \operatorname{Im} p^{(2)} \\ -\operatorname{Im} p^{(2)} & \operatorname{Re} p^{(2)} \end{pmatrix}.$$

Since $\Omega = x + y$, where Ω is the column $(\Omega_x, \Omega_y)^T$, we sum (7.3) and (7.4), which gives

$$\dot{\Omega} = Ax + BM + Cy + DM = A(\Omega - y) + BM + Cy + DM.$$

Relation(3.7) is equivalent to the following linear system of four equations:

$$\dot{z}(t) = \begin{pmatrix} \dot{\Omega} \\ \dot{y} \end{pmatrix} = \begin{pmatrix} A & -A + C \\ 0 & C \end{pmatrix} \begin{pmatrix} \Omega \\ y \end{pmatrix} + \begin{pmatrix} B + D \\ D \end{pmatrix} M. \tag{7.5}$$

Chapter 2

Control of a rotating rigid body containing a fluid with free surface

This Chapter is devoted to models describing rotational motions of a solid body with a cavity, one part of which is filled with an ideal fluid and the other part, with a gas at constant pressure.

Section 1 contains the setting of the problem of a rotating rigid body with a cavity filled with an ideal fluid with free surface. It is shown that, under the unperturbed motion, the free surface has cylindrical shape, provided that the problem is isoperimetric.

Section 2 considers small oscillations of a viscous fluid partially filling a rotating vessel. The problem is considered in the linear setting. By using Green's formula, integral equation is obtained, whose solution is taken for a generalized solution of the problem. Properties of the eigenvalues of the problem are also studied, and a variational statement is given.

The dynamics of the rotational motions of a rigid body with a cavity containing an ideal fluid with free surface is studied in Section 3. It is shown that the problem can be split into two parts, a hydrodynamic part, which describes the oscillations of a fluid in a fixed vessel, and a dynamical part, which describes the motion of a rigid body with a frozen fluid. These two problems can be solved independently.

Section 4 presents the linearized (with respect to the uniform rotation) initial boundary value problem on the perturbed motion of a rigid body with a cavity containing an ideal fluid with free surface.

Section 5 considers the hydrodynamic part of the problem; for a cavity of special shape, the eigenfunctions of the problem in the absence of perturbations are found by the method of separation of variables. It is shown that the eigenfunctions are orthogonal in the region occupied by the fluid.

In Section 6, the perturbed problem is solved by the Bubnov–Galerkin method. The hydrodynamic part of the problem is reduced to an infinite system of ordinary differential equations for the Fourier coefficients, and the motion of a rigid body with a fluid is described by ordinary differential equations for the vector of perturbations, whose coefficients reflect the influence of the wave motions of the fluid on the motion of the rigid body.

Section 7 studies the stability of the free rotation of the body–fluid system. For an axially symmetric body, the characteristic equation is given. Graphs of the coefficients in the inertial couplings as functions of the numbers l and p are constructed. A criterion for the stability of the steady rotation in the linear approximation is given.

In Section 8, the equations of motion are written in an equivalent form. By using the Laplace transform, an operator equation for the perturbed angular velocity is obtained, which is reduced with the use of convolution theorems to an integral relation relating the angular velocity to the external (control) moment. This integral relation is reduced to a system of ordinary differential equations.

1 STATEMENT OF THE PROBLEM

Consider a rigid body with a cavity D, one part of which is filled with an ideal fluid of density ρ and the other part, with a gas at pressure $p_0 = \text{const}$. We use the following notation: Q is the part of the cavity D occupied by the fluid; S is the wetted surface of the cavity; Σ is the free surface; $\boldsymbol{n}$ is the unit outer normal to the surface $S + \Sigma$ (see Fig. 2.1).

Suppose that the rigid body rotates about the Oz-axis with angular velocity $\boldsymbol{\omega}$. The fluid is in a field of mass forces with potential U. The equations of motion of the fluid in the body coordinate system $Oxyz$ have the form

$$\begin{cases} \boldsymbol{W}_0 + \boldsymbol{\omega} \times (\boldsymbol{\omega} \times \boldsymbol{R}) + \dot{\boldsymbol{\omega}} \times \boldsymbol{R} + 2\,\boldsymbol{\omega} \times \boldsymbol{v} + \dfrac{\partial \boldsymbol{v}}{\partial t} + (\boldsymbol{v}, \nabla)\boldsymbol{v} = -\dfrac{\nabla P}{\rho} - \nabla U, \\ \operatorname{div} \boldsymbol{v} = 0 \quad \text{on } Q, \quad (\boldsymbol{v}, \boldsymbol{n}) = 0 \quad \text{on } S, \\ (\boldsymbol{v}, \boldsymbol{n}) = -\dfrac{\partial F}{\partial t}, \quad P = P_0 \quad \text{on } \Sigma. \end{cases} \tag{1.1}$$

Here $\boldsymbol{v}$ denotes the velocity of the ideal fluid, t denotes time, $\boldsymbol{R}$ is the radius vector from the point O, P is the pressure, $\boldsymbol{W}_0$ is the absolute acceleration of the point O,

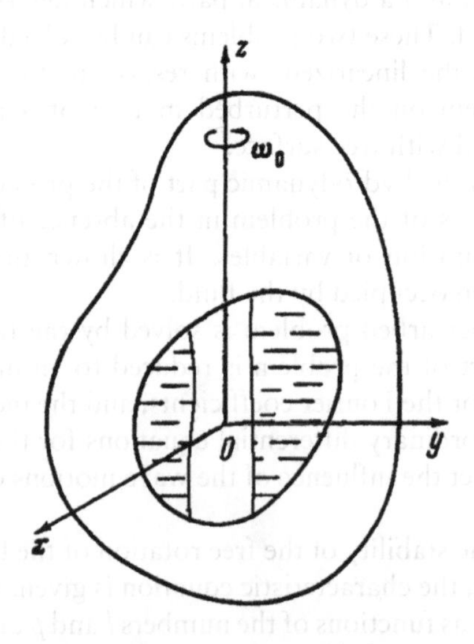

Figure 2.1 The body with a fluid.

ω is the absolute angular velocity of the body, $\dot{\omega}$ is its angular acceleration, P_0 is the gas pressure inside the cavity.

Consider the unperturbed motion of the fluid. We have

$$\omega = \omega_0 = \omega_0 e_z = \text{const}, \quad v = 0.$$

We set

$$\omega = \omega_0 + \Omega(t), \quad P = \rho[p - U - (w_0, r)], \quad M_1 = \omega_0 \times J\omega_0 + M \tag{1.2}$$

and assume that the quantities Ω, v, p, and M in the perturbed motion are first-order small. Substituting (1.2) into (1.1) and considering only the quantities which are first-order small, we reduce the equation of motion of the fluid to the form

$$\begin{cases} \dfrac{\partial v}{\partial t} + 2\,\omega_0 \times v = -\nabla p - \omega_0 \times (\omega_0 \times R) - \dot{\Omega} \times R, \\ \operatorname{div} v = 0 \quad \text{on } Q, \quad (v, n) = 0 \quad \text{on } S, \\ (v, n) = -\dfrac{\partial F}{\partial t}, \quad p = p_0 \quad \text{on } \Sigma. \end{cases} \tag{1.3}$$

Let us show that, under the unperturbed motion, the free surface has cylindrical shape.

Setting $v = 0$ and $\dot{\Omega} = 0$ in (1.3), we obtain

$$-\nabla p - \omega_0 \times (\omega_0 \times R) = 0.$$

Let us satisfy this equation and the boundary conditions. In the cylindrical coordinate system (e_r, e_φ, e_z), we have

$$\omega_0 = \omega_0 e_z, \quad R = re_r + ze_z,$$

$$\omega_0 \times (\omega_0 \times R) = \omega_0(\omega_0, R) - R\omega_0^2 = \omega_0^2(ze_z - re_r - ze_z) = \omega_0^2 re_r.$$

Next,

$$-\nabla p - \omega_0^2 re_r = 0, \quad \nabla\left(-p + \frac{\omega_0^2 r^2}{2}\right) = 0,$$

whence

$$p = p_0 + \frac{\omega_0^2 r^2}{2} + C.$$

The boundary conditions imply that, on Σ, $p = p_0$ and $C = -\omega_0^2 R_\Sigma^2/2$, where R_Σ is the radius of the free surface. Hence

$$p = p_0 + \frac{1}{2}\omega_0^2(r^2 - R_\Sigma^2).$$

Thus, under the unperturbed motion, the free surface of the rotating fluid is described by the equation $r = R_\Sigma$, where R_Σ is uniquely determined by the fluid mass $m = \int_Q \rho \, dV$.

On Σ, one more equation must be satisfied, namely, $(\partial F/\partial t) = 0$.

This can be achieved by setting $F = R_\Sigma$; thus, the system of equations (1.3) for the unperturbed motion is completely satisfied.

2 SMALL OSCILLATIONS OF A VISCOUS FLUID PARTIALLY FILLING A VESSEL

In this section, we study small oscillations of a fluid in a rotating vessel with a cavity partially filled with a viscous fluid. We establish some properties of the free oscillations of the rotating fluid and give the variational setting of the problem.

Statement of the problem. Suppose that a viscous incompressible fluid of mass m and kinematic viscosity ν partially fills a vessel of finite size, which rotates with angular velocity $\omega_0 = \text{const}$.

Let us introduce a coordinate system $Ox_1x_2x_3$ attached to the vessel so that the Ox_3-axis is directed along the angular velocity vector $\boldsymbol{\omega} = \omega_0 \boldsymbol{e}_3$. The equations of motion of the fluid and the boundary conditions in the system $Ox_1x_2x_3$ have the form

$$\begin{cases} \dfrac{\partial \boldsymbol{v}}{\partial t} + (\boldsymbol{v}, \nabla)\boldsymbol{v} + 2\boldsymbol{\omega} \times \boldsymbol{v} + \boldsymbol{\omega} \times (\boldsymbol{\omega} \times \boldsymbol{r}) = -\dfrac{\nabla P}{\rho} + \nabla U + \nu \Delta \boldsymbol{v}, \\ \operatorname{div} \boldsymbol{v} = 0 \quad \text{on } D, \\ \boldsymbol{v} = 0 \quad (\text{in the case of an ideal fluid, } (\boldsymbol{v}, \boldsymbol{n}) = 0) \quad \text{on } S, \\ \dfrac{d\Phi}{dt} = 0 \quad \text{when } \Phi(x_1, x_2, x_3, t) = 0, \text{ i.e., on } \Sigma. \end{cases} \tag{2.1}$$

Here S denotes the wetted surface of the cavity walls, Σ is the free surface of the fluid, $\boldsymbol{v}$ is the relative velocity of fluid particles in the system $Ox_1x_2x_3$, ρ is the density of the fluid, and U is the potential of the mass forces (for the gravitational force, we have $U = -gx_3$). The equation $\Phi(x_1, x_2, x_3, t) = 0$ described the free surface of the fluid.

Equations (2.1) are satisfied if

$$\begin{aligned} &\boldsymbol{v} = 0, \quad P = \rho\left[U + \frac{1}{2}(\boldsymbol{\omega} \times \boldsymbol{r})^2 + C_0\right] + P_0, \\ &\Phi = \Phi_0(x_1, x_2, x_3) = -U - \frac{1}{2}\omega_0^2\left(x_1^2 + x_2^2\right) - C_0. \end{aligned} \tag{2.2}$$

The constant C_0 is determined from the volume constancy condition (which means that the problem is isoperimetric)

$$\frac{m}{\rho} = \int_D dD, \tag{2.3}$$

where D is the volume bounded by the surfaces of the cavity walls and by the free surface $\Phi_0 = 0$. Solution (2.2) describes an equilibrium of the fluid in the system $Ox_1x_2x_3$. Consider small oscillations of the fluid about this equilibrium position.

Let

$$\Phi = \Phi_0 + |\nabla\Phi_0|\psi(x_1, x_2, x_3, t) = 0, \quad P = P_{st} + p, \tag{2.4}$$

where ψ and p are first-order small,

$$\frac{d\Phi}{dt} = \frac{\partial\Phi}{\partial t} + (v, \nabla)\Phi = |\nabla\Phi_0|\frac{\partial\psi}{\partial t} + (v, \nabla)\Phi = 0 \quad \text{when } \Phi = 0. \tag{2.5}$$

Up to terms of higher order of smallness, we obtain the following problem in linear setting:

$$\begin{cases} \dfrac{\partial v}{\partial t} + 2\omega \times v = -\dfrac{\nabla p}{\rho} + \nu\Delta v, \\ \operatorname{div} v = 0 \quad \text{on } D, \\ v = 0 \quad (\text{in the case of an ideal fluid } (v, n) = 0) \quad \text{on } S. \end{cases} \tag{2.6}$$

We use the notation $(\partial v_i/\partial x_k) = v_{i,k}$. Consider "normal" oscillations. Suppose that the time dependence of the velocity and pressure magnitudes (which we denote by the same letters as the velocity and pressure themselves) is determined by the factor $e^{\lambda t}$. We have

$$\begin{cases} \lambda v + 2\omega \times v = -\nu\Delta v + \nabla p, \\ \operatorname{div} v = 0 \quad \text{on } D, \quad v = 0 \quad \text{on } S, \\ v_{1,3} + v_{3,1} = 0, \quad v_{2,3} + v_{3,2} = 0, \\ p - 2\nu v_{3,3} = \dfrac{a}{\rho\lambda} v_n \quad \text{on } \Phi = 0; \end{cases} \tag{2.7}$$

the values of p and v are generally complex. In what follows, we use Green's formula, which is valid for sufficiently smooth solenoidal functions v and u with $u|_S = 0$. This formula is well known:

$$\int_\Omega (-\nu\Delta v + \nabla p)u^* d\Omega$$

$$= \nu E(v, u) - \int_\Sigma \left[\nu(v_{1,3} + v_{3,1})u_n^* + \nu(v_{2,3} + v_{3,2})u_n^* + (2\nu v_{3,3} - p)u_n^*\right] d\Sigma.$$

Here $E = \frac{1}{2}\int_\Omega \sum_{i,k=1}^{3} \left(\frac{\partial v_i}{\partial x_k} + \frac{\partial v_k}{\partial x_i}\right)\left(\frac{\partial u_i^*}{\partial x_k} + \frac{\partial u_k^*}{\partial x_i}\right) d\Omega$ is a bilinear functional with respect to the vector functions v and u^*; x_1, x_2, and x_3 are Cartesian coordinates; and the asterisk denotes complex conjugation. For the velocity vector v, the functional $E(v, v)$ is proportional to the energy dissipation in the volume.

Using Green's formula (and assuming the solution to be smooth enough), we obtain the integral relation

$$\lambda^2\rho\int_\Omega (\boldsymbol{v},\boldsymbol{u}^*)d\Omega + \int_\Sigma a v_n u_n^* d\Sigma - \lambda\rho\nu E(\boldsymbol{v},\boldsymbol{u}) + 2\omega_0\lambda\rho\int_\Omega \boldsymbol{e}_3(\boldsymbol{v}\times\boldsymbol{u}^*)d\Omega = 0. \quad (2.8)$$

We take relation (2.8) for the definition of a generalized solution of problem (2.7). We must seek the solution in the space of solenoidal vector functions from $W_2^1(\Omega)$ which vanish on S and have a normal derivative from $W_2^1(\Sigma)$ on the boundary Σ.

Relation (2.8) is equivalent to the requirement that the quadratic functional

$$J = \lambda^2\rho(\boldsymbol{v},\boldsymbol{v}) - i2\omega_0\lambda\rho C(\boldsymbol{v},\boldsymbol{v}) - \lambda\rho\nu E(\boldsymbol{v},\boldsymbol{v}) + B(v_n,v_n) \quad (2.9)$$

takes the zero stationary value at the solution of problem (2.7).

Here we use the notation

$$B(v_n,v_n) = \int_\Sigma a|v_n|^2 d\Sigma, \quad (\boldsymbol{v},\boldsymbol{v}) = \int_\Omega |\boldsymbol{v}|^2 d\Omega,$$

$$C(\boldsymbol{v},\boldsymbol{v}) = 2\int_\Omega \boldsymbol{e}_3(\mathrm{Re}\,\boldsymbol{v}\times\mathrm{Im}\,\boldsymbol{v})d\Omega,$$

$$\boldsymbol{e}_3(\boldsymbol{v}\times\boldsymbol{v}^*) = -2i\boldsymbol{e}_3(\mathrm{Re}\,\boldsymbol{v}\times\mathrm{Im}\,\boldsymbol{v}).$$

The problem can be approximately solved by Ritz' method. For the system of coordinate functions we can take any basis in the space of solenoidal vector functions with components from $W_2^1(\Omega)$ vanishing on S.

The limit cases. If $\boldsymbol{\omega} = 0$, then the functional under consideration coincides with the functional considered in [228], and if

$$\boldsymbol{\omega} = 0, \quad \nu = 0, \quad \boldsymbol{v} = \nabla\varphi, \quad \boldsymbol{\Delta}\varphi = 0, \quad U = -gx_3,$$

then it coincides with the well-known functional for oscillations of an ideal fluid [315].

Properties of the eigenvalues. Let $\boldsymbol{v}$ be an eigenfunction of problem (2.7); then

$$\lambda^2(\boldsymbol{v},\boldsymbol{v}) - i2\omega_0\lambda C(\boldsymbol{v},\boldsymbol{v}) - \lambda\nu E(\boldsymbol{v},\boldsymbol{v}) + \frac{1}{\rho}B(v_n,v_n) = 0. \quad (2.10)$$

In what follows, we use the inequalities

$$\begin{aligned} &E(\boldsymbol{v},\boldsymbol{v}) \geqslant \gamma(\boldsymbol{v},\boldsymbol{v}), \quad \gamma > 0, \quad C(\boldsymbol{v},\boldsymbol{v}) \leqslant (\boldsymbol{v},\boldsymbol{v}), \quad \text{because} \\ &2|\mathrm{Re}\,\boldsymbol{v}\times\mathrm{Im}\,\boldsymbol{v}| \leqslant 2|\mathrm{Re}\,\boldsymbol{v}||\mathrm{Im}\,\boldsymbol{v}| \leqslant |\mathrm{Re}\,\boldsymbol{v}|^2 + |\mathrm{Im}\,\boldsymbol{v}|^2 = |\boldsymbol{v}|^2. \end{aligned} \quad (2.11)$$

Let $\lambda = \alpha + i\beta$.

Substituting λ into (2.10) and separating the real and imaginary parts, we obtain

$$(\alpha^2+\beta^2)(\boldsymbol{v},\boldsymbol{v})+2\omega_0\beta C(\boldsymbol{v},\boldsymbol{v})-\alpha\nu E(\boldsymbol{v},\boldsymbol{v})+\frac{1}{\rho}B(v_n,v_n)=0,$$

$$-2\alpha\beta(\boldsymbol{v},\boldsymbol{v})+2\omega_0\alpha C(\boldsymbol{v},\boldsymbol{v})+\beta\nu E(\boldsymbol{v},\boldsymbol{v})=0.$$

Consider the case $\nu=0$.
We have $\alpha[\omega_0 C(\boldsymbol{v},\boldsymbol{v})-\beta(\boldsymbol{v},\boldsymbol{v})]=0$, whence $\alpha=0$.
Substituting this into the first equation, we obtain

$$\beta^2(\boldsymbol{v},\boldsymbol{v})-2\omega_0\beta C(\boldsymbol{v},\boldsymbol{v})-\frac{1}{\rho}B(v_n,v_n)=0,$$

$$\beta_{1,2}=\frac{1}{(\boldsymbol{v},\boldsymbol{v})}\left[\omega_0 C(\boldsymbol{v},\boldsymbol{v})\pm\sqrt{\omega_0^2C^2(\boldsymbol{v},\boldsymbol{v})+\frac{1}{\rho}(\boldsymbol{v},\boldsymbol{v})B(v_n,v_n)}\right].$$

We have

$$|\beta|\leqslant\frac{|\omega_0 C(\boldsymbol{v},\boldsymbol{v})|+\sqrt{\omega_0^2C^2(\boldsymbol{v},\boldsymbol{v})+\frac{1}{\rho}(\boldsymbol{v},\boldsymbol{v})B(v_n,v_n)}}{(\boldsymbol{v},\boldsymbol{v})},$$

$$|\beta|\leqslant\frac{|\omega_0 C(\boldsymbol{v},\boldsymbol{v})|+|\omega_0 C(\boldsymbol{v},\boldsymbol{v})|}{(\boldsymbol{v},\boldsymbol{v})}\leqslant 2\omega_0$$

(we have used the second inequality in (2.11)). For $\nu\neq 0$, we obtain $\operatorname{Re}\lambda\leqslant-\frac{1}{2}\nu\gamma$ and $|\operatorname{Im}\lambda|\leqslant 2\omega_0$ in a similar way.

Suppose that λ is a fixed number inside the interval $|\lambda|\leqslant 2\omega_0$, which contains all eigenvalues of problem (2.7). First, consider the case $\nu=0$. The equations and the boundary condition have the form

$$\begin{cases}\lambda\boldsymbol{v}+2\boldsymbol{\omega}\times\boldsymbol{v}=-\nabla p, \quad \operatorname{div}\boldsymbol{v}=0 \quad \text{on } D,\\ v_n=0 \quad \text{on } S, \quad p=\dfrac{a}{\lambda}v_n \quad \text{on } \Sigma.\end{cases} \tag{2.12}$$

Let us solve the first equation in (2.12) with respect to $\boldsymbol{v}$. To this end, we take its inner and vector product (on the left) with $\boldsymbol{\omega}_0$:

$$\begin{aligned}&\lambda(\boldsymbol{\omega}_0,\boldsymbol{v})+(\boldsymbol{\omega}_0,\nabla p)=0,\\ &2\boldsymbol{\omega}_0(\boldsymbol{\omega}_0,\boldsymbol{v})-2\omega_0^2\boldsymbol{v}+\lambda\boldsymbol{\omega}_0\times\boldsymbol{v}+\boldsymbol{\omega}_0\times\nabla p=0.\end{aligned} \tag{2.13}$$

Substitute the product $(\boldsymbol{\omega}_0,\boldsymbol{v})$ from the first relation in (2.13) and $\boldsymbol{\omega}_0\times\boldsymbol{v}$ from the first equation in (2.12) into the second equation in (2.13) and solving this equation with respect to $\boldsymbol{v}$, we obtain

$$\boldsymbol{v}=\frac{2\boldsymbol{\omega}_0\times\nabla p-\lambda\nabla p-4\lambda^{-1}\boldsymbol{\omega}_0(\boldsymbol{\omega}_0,\nabla p)}{\lambda^2+4\omega_0^2}. \tag{2.14}$$

Following [36], we set $\sigma = 2\omega_0/\lambda$ (this is a complex number) and introduce a linear transformation L by setting

$$La = a + \sigma^2 e_3(e_3, a) + \sigma\, a \times e_3,$$

$$L = \begin{pmatrix} 1 & \sigma & 0 \\ -\sigma & 1 & 0 \\ 0 & 0 & 1+\sigma^2 \end{pmatrix}; \tag{2.15}$$

here a is an arbitrary vector. The matrix L determines a linear transformation in the coordinate system $Ox_1x_2x_3$ and has the properties

$$L'(\sigma) = L(-\sigma), \quad (a, Lb) = (L'a, b).$$

Here the prime denotes matrix transposition. Using the matrix L, we write the expression for the velocity and the problem for determining p in the form

$$\begin{cases} v = -\lambda^{-1}(1+\sigma^2)^{-1} L\nabla p, \quad \operatorname{div} L\nabla p = 0 \quad \text{on } D, \\ (L\nabla p, n) = 0 \quad \text{on } S, \quad p = \dfrac{a}{\lambda}(L\nabla p, n) \quad \text{on } \Sigma. \end{cases} \tag{2.16}$$

We seek a solution of (2.16) in the form

$$p = -\lambda\varphi, \quad v = \frac{L\nabla\varphi}{1+\sigma^2}, \tag{2.17}$$

where φ is an unknown function. We arrive at the following problem for the function φ:

$$\begin{cases} \Delta\varphi + \sigma^2 \dfrac{\partial^2\varphi}{\partial x_3^2} = 0 \quad \text{on } D, \\ (L\nabla\varphi, n) = 0 \quad \text{on } S, \\ \lambda^2\varphi + \dfrac{a}{1+\sigma^2}(L\nabla\varphi, n) = 0 \quad \text{on } \Sigma. \end{cases} \tag{2.18}$$

Variational principle. Using relation (2.9) obtained in this chapter, we can easily show that problem (2.18) admits the following variational setting:

$$\delta\left(\lambda^2 \int_\Omega (L\nabla\varphi)^2 d\Omega + 2\omega_0\lambda \int_\Omega [(L\nabla\varphi) \times (L\nabla\varphi^*)] d\Omega + \int_\Sigma (L\nabla\varphi, n)^2 a\, d\Sigma\right) = 0.$$

Setting $\sigma_1 = 2\omega_0/\lambda = i\sigma$ and $\omega_0 \neq 0$, we obtain

$$\delta\left(\int_\Omega (L\nabla\varphi)^2 d\Omega - i\sigma_1 \int_\Omega [(L\nabla\varphi) \times (L\nabla\varphi^*)] d\Omega - \frac{4\omega_0^2}{\sigma_1^2} \int_\Sigma (L\nabla\varphi, n)^2 a\, d\Sigma\right) = 0. \tag{2.19}$$

The minimum is sought in the class of functions satisfying the equation

$$\Delta\varphi - \sigma_1^2 \frac{\partial^2 \varphi}{\partial x_3^2} = 0. \tag{2.20}$$

3 ROTATIONAL MOTIONS OF A RIGID BODY WITH A CAVITY PARTIALLY FILLED WITH A FLUID

Consider a rigid body with a cavity D one part of which is filled with an ideal fluid of density ρ and the other part, with a gas at pressure $p_0 = \text{const}$; the body is assumed to move in the field of mass forces with potential U. The domain Q occupied by the fluid is bounded by the wetted surface S of the cavity and a free surface Σ.

The equations of motion for the fluid and the boundary and initial conditions in an arbitrary coordinate system $Ox_1x_2x_3$ attached to the rigid body are

$$\begin{cases} \boldsymbol{w}_0 + \boldsymbol{\omega} \times (\boldsymbol{\omega} \times \boldsymbol{r}) + \dot{\boldsymbol{\omega}} \times \boldsymbol{r} + 2\boldsymbol{\omega} \times \boldsymbol{V} + \dfrac{\partial \boldsymbol{V}}{\partial t} + (\boldsymbol{V}, \nabla)\boldsymbol{V} = -\dfrac{\nabla P}{\rho} - \nabla U, \\ \operatorname{div} \boldsymbol{V} = 0 \quad \text{on } Q, \quad (\boldsymbol{V}, \boldsymbol{n}) = 0 \quad \text{on } S, \\ \dfrac{dF}{dt} = 0, \quad P = P_0 \quad \text{on } \Sigma, \quad \boldsymbol{V} = \boldsymbol{V}_0(\boldsymbol{r}) \quad \text{at } t = 0. \end{cases} \tag{3.1}$$

Here, and in what follows, the dot denotes differentiation with respect to time in the coordinate system $Ox_1x_2x_3$; $\boldsymbol{w}_0$ is the absolute acceleration of the point O, $\boldsymbol{\omega}$ is the absolute angular velocity of the body, $\dot{\boldsymbol{\omega}}$ is its angular acceleration, $\boldsymbol{r}$ is the radius vector from the point O, $\boldsymbol{V}$ is the velocity of the fluid in the coordinate system $Ox_1x_2x_3$, t is time, P is the pressure, $\boldsymbol{n}$ is the unit outer normal vector to $S+\Sigma$, and $F(x_1, x_2, x_3, t) = 0$ is the equation of the free surface of the fluid.

We write the kinetic momentum of the body with a fluid about the center of inertia O_1 of the entire system in the form

$$\boldsymbol{K} = J\boldsymbol{\omega} + \rho \int_{Q_0} \boldsymbol{r} \times \boldsymbol{V} dQ. \tag{3.2}$$

Here J is the inertia tensor of the entire system with respect to the point O_1, which is formed by the inertia tensors J^1 and J^2 of the body and the hardened fluid, respectively, relative to the same point. The domain Q_0 is bounded by the wetted surface S and the free surface Σ in the unperturbed motion, which has cylindrical shape in the first approximation [91]. Thus, the kinetic momentum of the body–fluid system equals the sum of the kinetic momentum of the absolutely rigid body, which is equivalent to that of the same system in which the free surface is replaced by a cylindrical lid, and the kinetic momentum of the wave motion of the fluid.

Thus, in the approximation under consideration, the body with a fluid-containing cavity is a gyrostat; therefore, the inertia center O_1 of the system is fixed with respect to the coordinate system $Ox_1x_2x_3$, and the tensors J^1, J^2, and J are constant in this coordinate system. The second term in (3.2), which is called the gyrostatic moment, does not depend on the choice of the pole [36] and can be calculated with respect to the point O; this is done in (3.2).

Let us write the equation for the moments about the point O_1 in the body coordinate system $Ox_1x_2x_3$:

$$\dot{K} + \omega \times K = M_1. \tag{3.3}$$

Here M_1 is the resultant moment about O_1 of all external forces acting on the body with a fluid.

Equations (3.1)–(3.3), together with the usual equations of motion for the center of inertia, the kinematic relations, and the initial conditions completely describe the dynamics of the body with a fluid.

4 LINEARIZATION OF THE PROBLEM

Suppose that the unperturbed motion of the body with a fluid with respect to the center of inertia O_1 consists in the rotation of the entire system about the axis O_1y, which passes through O_1 and is parallel to the Ox_3 axis, with constant angular velocity ω_0 so that the free surface has cylindrical shape. For the unperturbed motion, we have

$$\omega = \omega_0 = \omega_0 e_3, \quad V = 0, \quad M_1 = \omega_0 \times J\omega_0,$$
$$F(r, \theta, z) = r_0 - r = 0,$$

where e_3 the unit vector along the Ox_3 axis, r_0 is the radius of the free surface in the unperturbed motion, $x_1 = r\cos\theta$, $x_2 = r\sin\theta$, and $x_3 = z$.

Consider the perturbed motion of the system. We set

$$\begin{aligned} &\omega(t) = \omega_0 + \Omega(t), \quad P = \rho\left[p + \frac{1}{2}(\omega \times r)^2 - U - (w_0, r)\right], \\ &M_1 = \omega_0 \times J\omega_0 + M, \quad F(r, \theta, z, t) = r_0 - r + h(\theta, z, t) = 0 \end{aligned} \tag{4.1}$$

and assume the quantities Ω, V, M, p, and h to be first-order small in the perturbed motion.

Substituting (4.1) into equation (3.1) and discarding terms of higher order of smallness, we reduce the problem of the motion of the fluid to the form

$$\begin{cases} \dfrac{\partial V}{\partial t} + 2\,\omega_0 \times V + \dot{\Omega} \times r = -\nabla p, \\ \operatorname{div} V = 0 \quad \text{on } Q, \quad (V, n) = 0 \quad \text{on } S, \\ \dfrac{\partial p}{\partial t} = -\left(\omega_0 \times V + \dot{\Omega} \times r, \omega_0 \times r\right), \\ \dfrac{\partial h}{\partial t} = -(V, n) \quad \text{on } \Sigma, \quad V = V_0(r) \quad \text{at } t = 0. \end{cases} \tag{4.2}$$

Similarly, the equations of motion for the body with a fluid take the form

$$J\dot{\boldsymbol{\Omega}} + \boldsymbol{\Omega} \times J\boldsymbol{\omega}_0 + \boldsymbol{\omega}_0 \times J\boldsymbol{\Omega} + \rho \int_Q \boldsymbol{r} \times \dot{\boldsymbol{V}} dQ + \rho \int_Q \boldsymbol{\omega}_0 \times (\boldsymbol{r} \times \boldsymbol{V}) dQ = \boldsymbol{M}. \tag{4.3}$$

5 HYDRODYNAMIC PROBLEM

Let us set $\boldsymbol{\Omega} = 0$ in problem (4.2) and consider the auxiliary problem on the oscillations of a fluid in a uniformly rotating vessel; we shall seek the solution in the harmonic oscillation form

$$\boldsymbol{V} = \boldsymbol{u}(x_1, x_2, x_3)e^{\lambda t}, \quad p = \varphi(x_1, x_2, x_3)e^{\lambda t}, \quad h = f(\theta, z)e^{\lambda t}.$$

Thus, we have the following problem:

$$\begin{cases} \lambda \boldsymbol{u} + 2\,\boldsymbol{\omega}_0 \times \boldsymbol{u} + \nabla\varphi = 0, \\ \operatorname{div} \boldsymbol{u} = 0 \quad \text{on } Q, \quad (\boldsymbol{u}, \boldsymbol{n}) = 0 \quad \text{on } S, \\ \lambda\varphi = -(\boldsymbol{\omega}_0 \times \boldsymbol{u}, \boldsymbol{\omega}_0 \times \boldsymbol{r}), \\ \lambda f = -(\boldsymbol{u}, \boldsymbol{n}) \quad \text{on } \Sigma. \end{cases} \tag{5.1}$$

Apply the linear transformation $L(\sigma)$ [57] defined by

$$L\boldsymbol{b} = \boldsymbol{b} + \sigma^2 \boldsymbol{e}_3(\boldsymbol{e}_3, \boldsymbol{b}) + \sigma\, \boldsymbol{b} \times \boldsymbol{e}_3,$$

we obtain the following boundary eigenvalue problem for the function $\varphi(x_1, x_2, x_3)$:

$$\begin{cases} \Delta\varphi + \sigma^2 \dfrac{\partial^2 \varphi}{\partial x_3^2} = 0 \quad \text{on } Q, \\ (L\nabla\varphi, \boldsymbol{n}) = 0 \quad \text{on } S, \\ \lambda^2 \varphi + \dfrac{r_0}{1+\sigma^2}(L\nabla\varphi, \boldsymbol{n}) = 0, \\ (L\nabla\varphi, \boldsymbol{n}) = \lambda^2(1+\sigma^2) f \quad \text{on } \Sigma. \end{cases} \tag{5.2}$$

The eigenfunctions φ_n $(n = 1, 2, \ldots)$ of the boundary value problem (5.2) correspond to the eigenvalues $\sigma_n = 2\omega_0/\lambda_n$, which densely fill the interval $|\sigma_n| \geqslant 1$ of the imaginary axis [91]. For practically interesting shapes of the cavity D, the system of functions φ_n turns out to be complete and orthogonal.

Hereafter we consider a dynamically symmetric body with a cylindrical cavity D and a fluid occupying the domain

$$Q = \left\{(r, \theta, z) : r_0 \leqslant r \leqslant 1, 0 \leqslant \theta < 2\pi, -h \leqslant z \leqslant h\right\}.$$

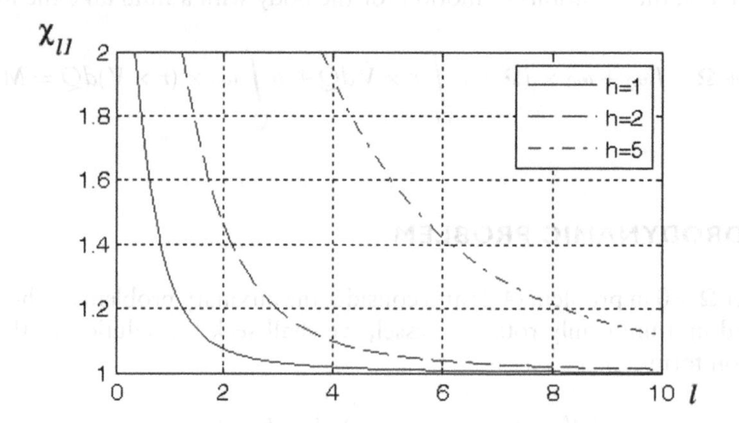

Figure 2.2 The frequencies χ_{ll} at $r_0 = 0.8$.

In this case, the solution of problem (5.2) written in the cylindrical coordinate system ($x_1 = r\cos\theta$, $x_2 = r\sin\theta$, $x_3 = z$) has the form

$$\varphi_n = \varphi_{lp}(r,\theta,z) = [J_1(\xi_{lp}r) + C_{lp}Y_1(\xi_{lp}r)]\sin(k_l z)e^{i\theta},$$

$$C_{lp} = \frac{J_1'(\xi_{lp}) + i\sigma_{lp}J_1(\xi_{lp})}{Y_1'(\xi_{lp}) + i\sigma_{lp}Y_1(\xi_{lp})}, \quad k_l = \pi(2l+1)/2h, \tag{5.3}$$

$$\xi_{lp} = ik_l\sqrt{\sigma_{lp}^2 + 1}, \quad l = 0,1,\ldots, \quad p = 1,2,\ldots.$$

The remaining harmonics are not excited in θ and z in this mechanical system. The index n ranges over all combinations of numbers of the longitudinal and transverse harmonics l and p. The quantity ξ_{lp} is the pth root of the equation

$$\begin{aligned}
&\left[J_1'(\xi_{lp}) + i\sigma_{lp}J_1(\xi_{lp})\right]\cdot\left[Y_1'(\xi_{lp}r_0) + K_{lp}Y_1(\xi_{lp}r_0)\right] \\
&\quad - \left[J_1'(\xi_{lp}r_0) + K_{lp}J_1(\xi_{lp}r_0)\right]\cdot\left[Y_1'(\xi_{lp}) + i\sigma_{lp}Y_1(\xi_{lp})\right] = 0, \\
&K_{lp} = \frac{i\sigma_{lp}^3 - 4(1+\sigma_{lp}^2)}{r_0\sigma_{lp}^2}.
\end{aligned}$$

Figures 2.2 and 2.3 present the graphs of the eigenfrequencies $\chi_{l1} = i\sigma_{l1}$ for a cylindrical cavity with height $2h$ and radius $r_1 = 1$ for the cases $r_0 = 0.8$ and $r_0 = 0.7$, respectively. Similarly, in Figs. 2.4 and 2.5, the graphs of χ_{0p} are shown.

The vector functions

$$u_n = -\lambda_n^{-1}(1+\sigma_n^2)^{-1}L\nabla\varphi_n \tag{5.4}$$

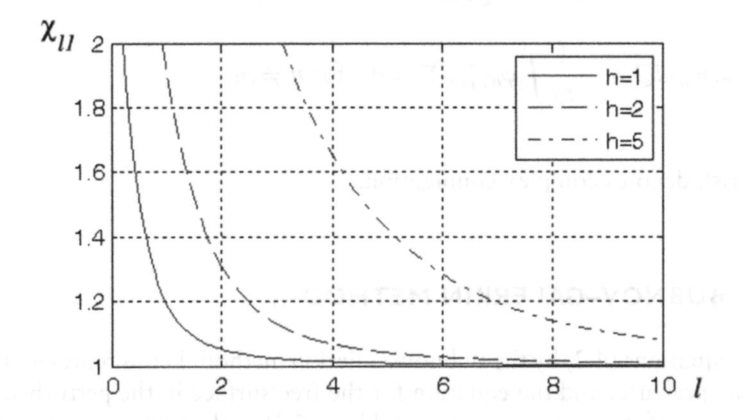

Figure 2.3 The frequencies χ_{ll} at $r_0 = 0.7$.

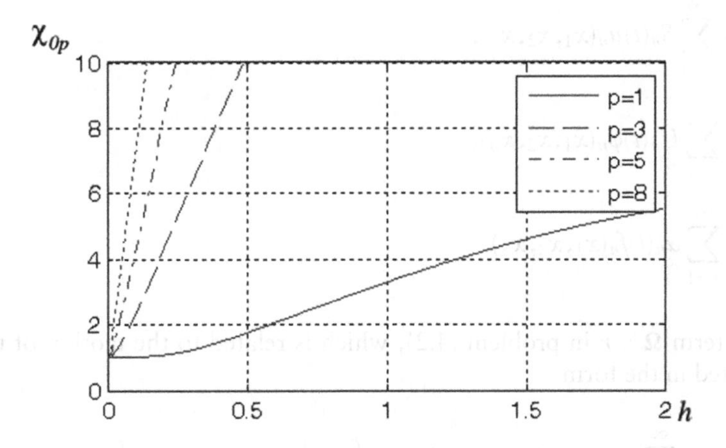

Figure 2.4 The frequencies χ_{0p} at $r_0 = 0.8$.

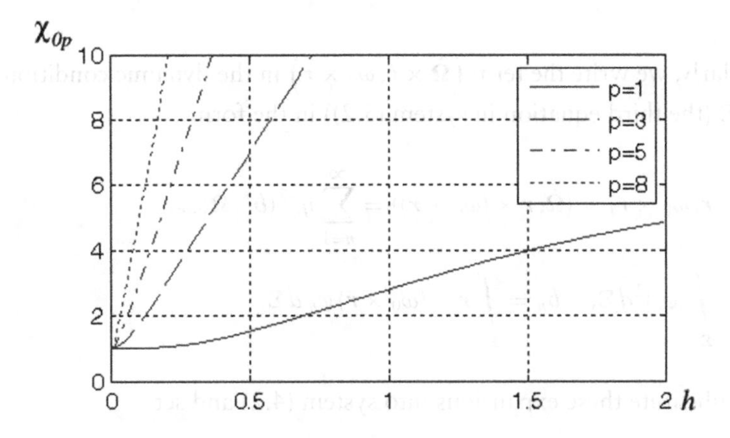

Figure 2.5 The frequencies χ_{0p} at $r_0 = 0.7$.

are orthogonal in the domain Q, that is,

$$\int_Q (\boldsymbol{u}_n, \boldsymbol{u}_m^*) dQ = -\frac{1}{r_0} \int_\Sigma \varphi_n \varphi_m^* \, d\Sigma = 0 \quad \text{for } n \neq m. \tag{5.5}$$

The asterisk denotes complex conjugation.

6 THE BUBNOV–GALERKIN METHOD

We solve equations (4.2) by the Bubnov–Galerkin method. Let us represent the velocity vector, the pressure, and the equation for the free surface in the perturbed motion as the expansion of the boundary value problem (5.2) in the vector eigenfunctions (5.5):

$$\boldsymbol{V} = \sum_{n=1}^{\infty} S_n(t) \boldsymbol{u}_n(x_1, x_2, x_3),$$

$$p = \sum_{n=1}^{\infty} U_n(t) \varphi_n(x_1, x_2, x_3),$$

$$h = \sum_{n=1}^{\infty} \alpha_n(t) f_n(x_1, x_2, x_3).$$

The term $\dot{\boldsymbol{\Omega}} \times \boldsymbol{r}$ in problem (4.2), which is related to the motion of the body, is represented in the form

$$\dot{\boldsymbol{\Omega}} \times \boldsymbol{r} = \sum_{n=1}^{\infty} \mu_n^{-2} (\boldsymbol{a}_n^*, \dot{\boldsymbol{\Omega}}) \boldsymbol{u}_n, \quad \mu_n^2 = \rho \int_Q |\boldsymbol{u}_n|^2 dQ, \quad \boldsymbol{a}_n = \rho \int_Q \boldsymbol{r} \times \boldsymbol{u}_n \, dQ.$$

Similarly, we write the term $\left(\dot{\boldsymbol{\Omega}} \times \boldsymbol{r}, \boldsymbol{\omega}_0 \times \boldsymbol{r}\right)$ in the dynamic condition on the free surface Σ (the third equation in system (5.2)) in the form

$$(\dot{\boldsymbol{\Omega}} \times \boldsymbol{r}, \boldsymbol{\omega}_0 \times \boldsymbol{r}) = (\dot{\boldsymbol{\Omega}}, \boldsymbol{r} \times (\boldsymbol{\omega}_0 \times \boldsymbol{r})) = \sum_{n=1}^{\infty} \eta_n^{-2} (\boldsymbol{b}_n^*, \dot{\boldsymbol{\Omega}}) \varphi_n,$$

$$\eta_n^2 = \int_\Sigma |\varphi_n|^2 d\Sigma, \quad \boldsymbol{b}_n = \int_\Sigma \boldsymbol{r} \times (\boldsymbol{\omega}_0 \times \boldsymbol{r}) \varphi_n \, d\Sigma.$$

We substitute these expansions into system (4.2) and set

$$2\,\boldsymbol{\omega}_0 \times \boldsymbol{u}_n = -\lambda_n \boldsymbol{u}_n - \nabla \varphi_n, \quad (\boldsymbol{u}_n, \boldsymbol{n}_\Sigma) = -\lambda_n f_n.$$

Next, we take the inner product of the first equation in the system with $\boldsymbol{u}_m^*$ and of the third and the fourth equations (the conditions on the free surface) with φ_m^* and f_m^*, respectively; then, we integrate the first equation over the volume Q and the third and fourth ones over the surface Σ. By virtue of the orthogonality property (5.5), system (4.2) reduces to the following infinite system of ordinary differential equations for the coefficients S_n, U_n, and α_n:

$$\begin{aligned} &\dot{S}_n - \lambda_n[S_n - \delta_n(S_n - U_n)] + \mu_n^{-2}(\boldsymbol{a}_n^*, \dot{\boldsymbol{\Omega}}) = 0, \\ &\dot{U}_n - \lambda_n S_n + \eta_n^{-2}(\boldsymbol{b}_n^*, \dot{\boldsymbol{\Omega}}) = 0, \\ &\dot{\alpha}_n - \lambda_n S_n = 0, \quad \delta_n = \frac{\eta_n^2}{r_0 \mu_n^2} \quad (n = 1, 2, \ldots). \end{aligned} \tag{6.1}$$

The linearized equation of motion of the rigid body with the above expansions taken into account has the form

$$(J^0 + J^1)\dot{\boldsymbol{\Omega}} + \boldsymbol{\Omega} \times (J^0 + J^1)\boldsymbol{\omega}_0 + \boldsymbol{\omega}_0 \times (J^0 + J^1)\boldsymbol{\Omega} + \sum_{n=1}^{\infty} \left[\boldsymbol{a}_n \dot{S}_n + (\boldsymbol{\omega}_0 \times \boldsymbol{a}_n) S_n\right] = \boldsymbol{M}. \tag{6.2}$$

Here J^0 is the inertia tensor of the body without fluid, J^1 is the inertia tensor of the hardened fluid, and $\boldsymbol{M}$ is the moment of external forces.

The system of equations for the perturbed motion of the body in the form (6.1), (6.2) makes it possible to consider regimes in which the body is subject to the action of any control moments.

If the axis of rotation of the system in the unperturbed motion is simultaneously the geometrical axis of symmetry and the axis of mass symmetry, then the equations can be significantly simplified. For a dynamically symmetric body, the scalar equation of motion about the Ox_3 axis can be separated from the other equations. The equations with respect to the transverse axes Ox_1 and Ox_2 are identical. The components of the vector quantities $\boldsymbol{a}_n = (a_{1n}, a_{2n}, a_{3n})$ and $\boldsymbol{b}_n = (b_{1n}, b_{2n}, b_{3n})$ are related by

$$\begin{aligned} &a_{1n} = a_{1n}^* = -ia_{2n} = ia_{2n}^* = a_n, \\ &b_{1n} = b_{1n}^* = -ib_{2n} = ib_{2n}^* = b_n. \end{aligned}$$

In this case, equations (6.1) and (6.2) can be written in the form

$$\begin{aligned} &A\dot{\Omega} + i(C - A)\omega_0\Omega + \sum_{n=1}^{\infty} 2a_n(\dot{S}_n - i\omega_0 S_n) = M, \\ &\dot{S}_n - i\gamma_n[S_n - \delta_n(S_n - U_n)] + \mu_n^{-2} a_n \dot{\Omega} = 0, \\ &\dot{U}_n - i\gamma_n S_n + \eta_n^{-2} b_n \dot{\Omega} = 0, \\ &\dot{\alpha}_n - i\gamma_n S_n = 0, \quad \lambda_n = i\gamma_n \quad (n = 1, 2, \ldots). \end{aligned} \tag{6.3}$$

Here

$$A = J_{11}^0 + J_{11}^1 = J_{22}^0 + J_{22}^1, \quad C = J_{33}^0 + J_{33}^1,$$
$$\Omega = \Omega_1 - i\Omega_2, \quad M = M_1 - iM_2.$$

7 STABILITY OF THE FREE ROTATION OF A BODY–FLUID SYSTEM

Consider the stability of the free rotation of the system described above. The characteristic equation of system (6.3) at $M = 0$ has the form

$$Aq + (C - A)\omega_0 - q(q - \omega_0)\sum_{n=1}^{\infty}\left(\frac{H_n}{q - \gamma_n} + \frac{F_n}{q + \delta_n\gamma_n}\right) = 0, \tag{7.1}$$

$$H_n = \frac{2a_n(r_0 a_n + b_n)}{\eta_n^2 + r_0\mu_n^2}, \quad F_n = \frac{2a_n(r_0 a_n\delta_n - b_n)}{\eta_n^2 + r_0\mu_n^2}.$$

Before proceeding to study the characteristic equation (7.1), we must investigate the dependence of the behavior of the coefficients H_{lp} and F_{lp} in the inertial couplings on the numbers l and p. For this purpose, numerical algorithms were designed for constructing the corresponding dependences. The graphs of these dependences are shown in Figs. 2.6–2.13. The coupling coefficients δ_{lp} as functions of the numbers l and p were also studied. The results of calculations are presented in Figs. 2.14–2.17.

Figures 2.6 and 2.7 show the dependence of the inertial coupling coefficient H_{lp} on l at fixed $p = 1$ for a cylindrical cavity in the cases $r_0 = 0.8$ and $r_0 = 0.7$.

Figures 2.8 and 2.9 show the dependence of the inertial coupling coefficients H_{lp} on p at fixed $l = 0$ for a cylindrical cavity in the cases $r_0 = 0.8$ and $r_0 = 0.7$.

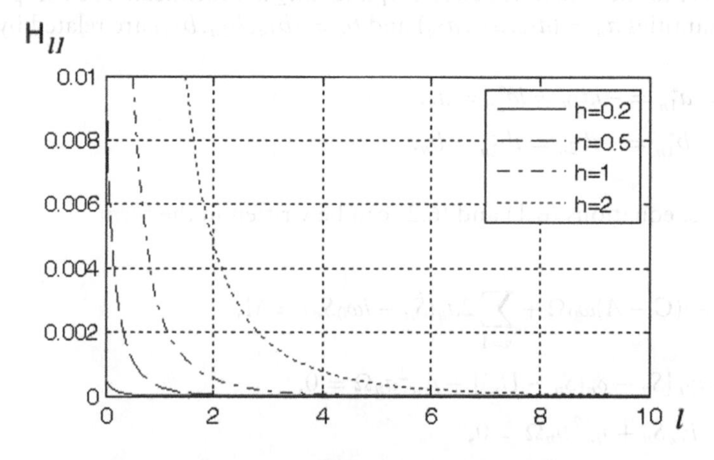

Figure 2.6 The coefficients H_{ll} under $r_0 = 0.8$.

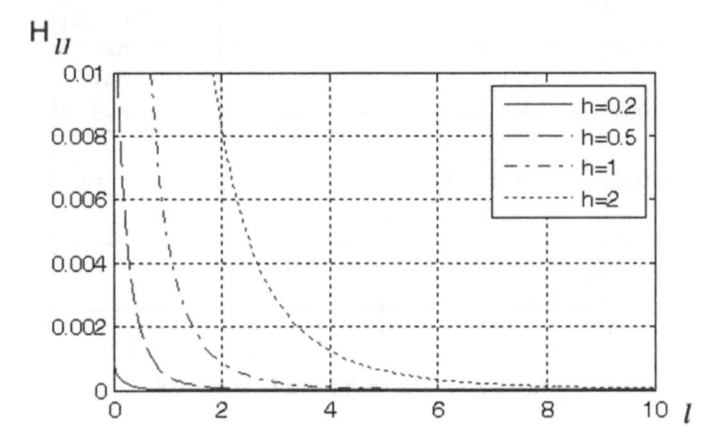

Figure 2.7 The coefficients H_{ll} at $r_0 = 0.7$.

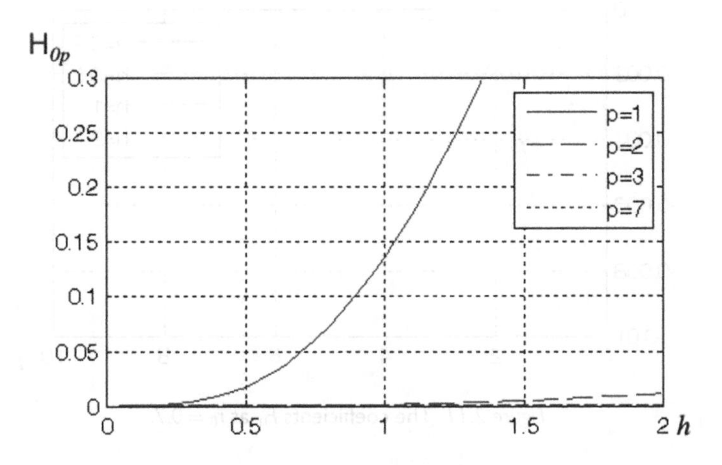

Figure 2.8 The coefficients H_{0p} under $r_0 = 0.8$.

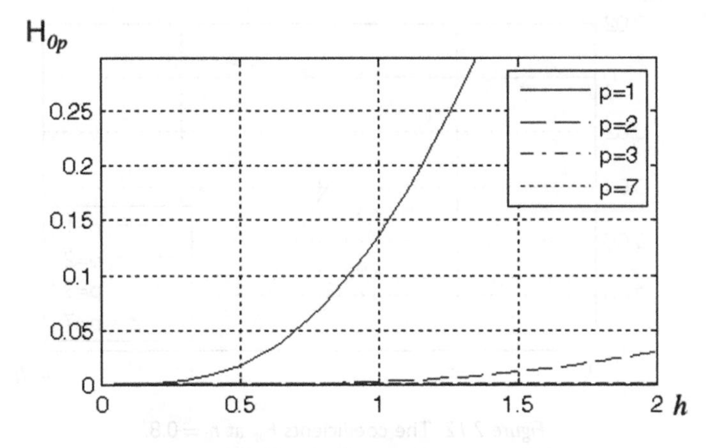

Figure 2.9 The coefficients H_{0p} at $r_0 = 0.7$.

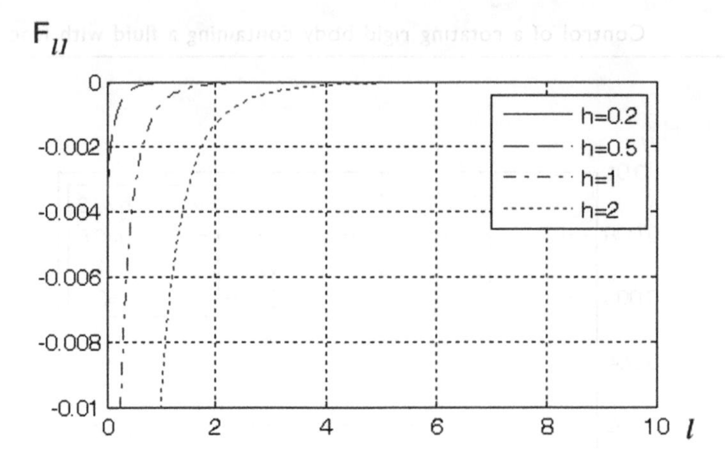

Figure 2.10 The coefficients F_{ll} at $r_0 = 0.8$.

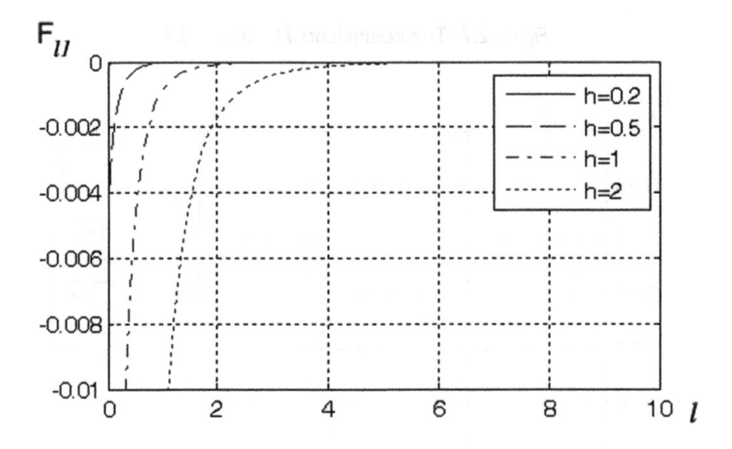

Figure 2.11 The coefficients F_{ll} at $r_0 = 0.7$.

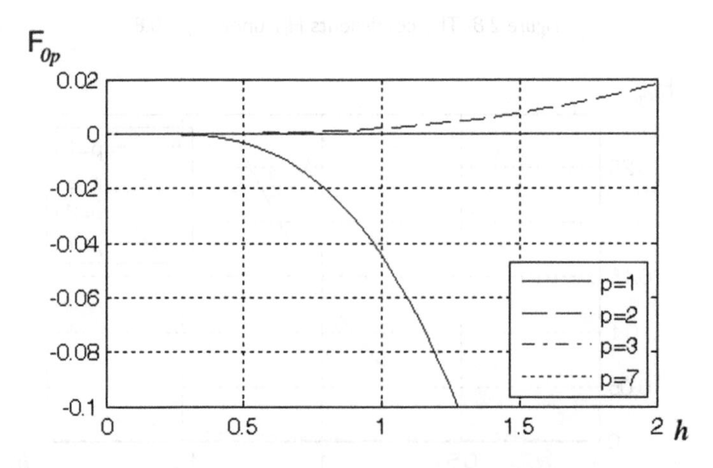

Figure 2.12 The coefficients F_{0p} at $r_0 = 0.8$.

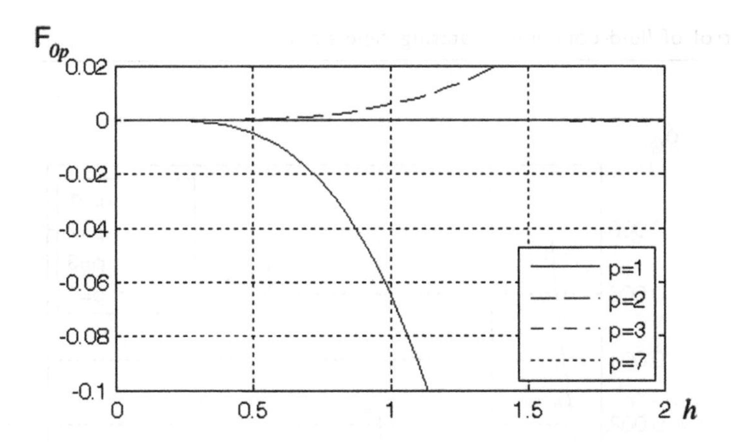

Figure 2.13 The coefficients F_{0p} at $r_0 = 0.7$.

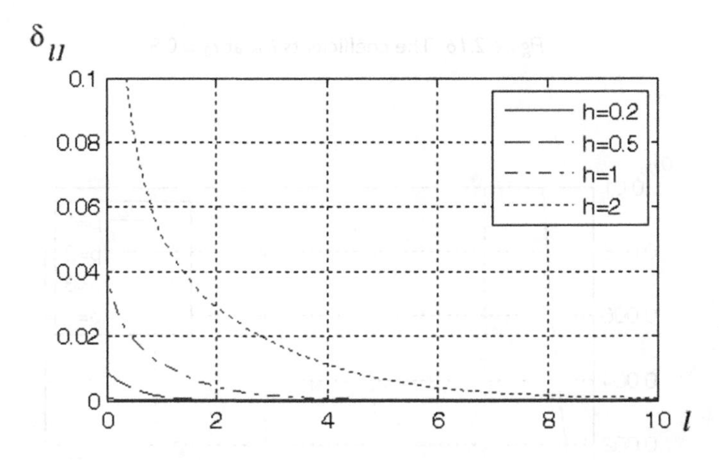

Figure 2.14 The coefficients δ_{ll} at $r_0 = 0.8$.

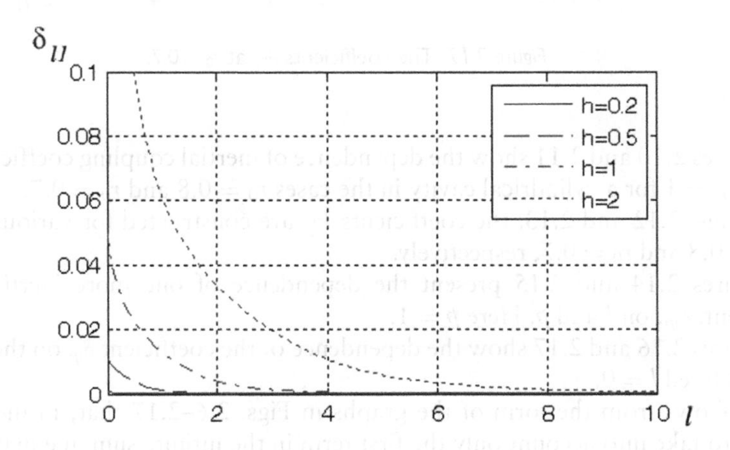

Figure 2.15 The coefficients δ_{ll} at $r_0 = 0.7$.

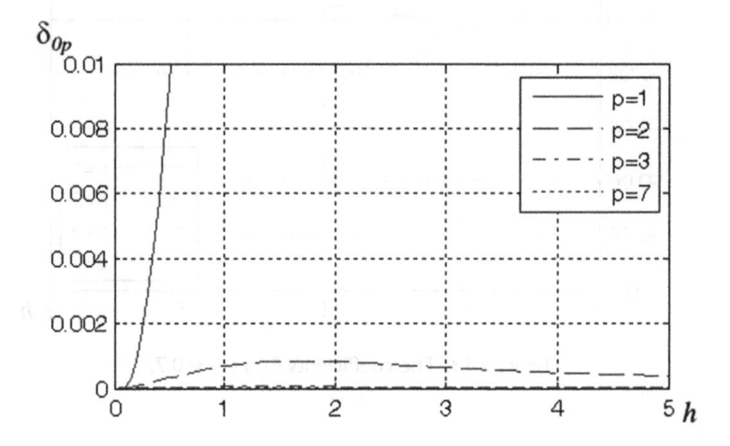

Figure 2.16 The coefficients δ_{0p} at $r_0 = 0.8$.

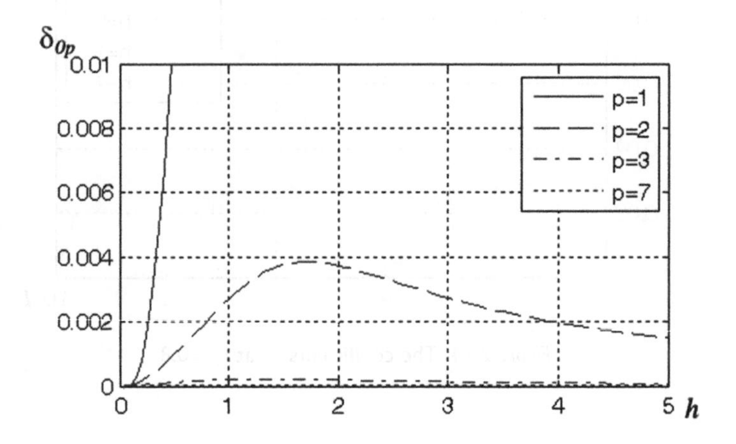

Figure 2.17 The coefficients δ_{0p} at $r_0 = 0.7$.

Figures 2.10 and 2.11 show the dependence of inertial coupling coefficient F_{lp} on l at fixed $p = 1$ for a cylindrical cavity in the cases $r_0 = 0.8$ and $r_0 = 0.7$.

In Figs. 2.12 and 2.13, the coefficients F_{lp} are constructed for various p at $l = 0$ for $r_0 = 0.8$ and $r_0 = 0.7$, respectively.

Figures 2.14 and 2.15 present the dependence of one more inertial coupling coefficient, δ_{lp}, on l and p. Here $p = 1$.

Figures 2.16 and 2.17 show the dependence of the coefficient δ_{lp} on the numbers l and p at fixed $l = 0$.

It follows from the form of the graphs in Figs. 2.6–2.17 that, in most cases, it suffices to take into account only the first term in the infinite sum in equation (7.1).

For a steady rotation to be stable, it is necessary that all roots of equation (7.1) be real. We shall consider only the first approximation, retaining only the principal term ($n = 1$) of the infinite sum in (7.1). In this case, the characteristic equation is a third-degree polynomial in q:

$$q^3 + z_2 q^2 + z_1 q + z_0 = 0. \tag{7.2}$$

Here

$$\begin{aligned}
z_2 &= z_3^{-1}[A\gamma_1(\delta_1 - 1) - \gamma_1(H_1\delta_1 - F_1) + (H_1 + F_1 + C - A)\omega_0],\\
z_1 &= z_3^{-1}\gamma_1[\omega_0((C - A)(\delta_1 - 1) + H_1\delta_1 - F_1) - A\gamma_1\delta_1],\\
z_0 &= z_3^{-1}(A - C)\gamma_1^2\delta_1\omega_0, \quad z_3 = A - F_1 - H_1.
\end{aligned}$$

Replacing the unknown q in equation (7.2) by the new unknown $\bar{q}$ related to q by

$$q = \bar{q} - \frac{z_2}{3},$$

we obtain the following incomplete cubic equation for $\bar{q}$:

$$\bar{q}^3 + x\bar{q} + y = 0,$$

where

$$x = z_1 - \frac{z_2^2}{3}, \quad y = z_0 - \frac{z_2}{3}\left(\frac{z_2^2}{9} + z_1\right).$$

For the free rotation to be stable, the parameters of the system must satisfy the condition [193]

$$D = -4x^3 - 27y^2 \geqslant 0.$$

On the boundary of the stability domain, the condition $D = 0$ holds.

This point needs explanation. The above stability condition was obtained from the characteristic equation (7.1) with only the first term of the infinite series left; thus, it does not look convincing. Moreover, Academician S. L. Sobolev mentioned in [299] that, under the variation of the rotation velocity of a cylinder partially filled with an ideal fluid, stability may be lost, then restored, and this may happen many times. In the case under consideration, this is not so. The general angular rate ω_0, with respect to which the equations of motion are linearized, is large in comparison with the perturbations Ω_x and Ω_y perpendicular to the main rotation, which are first-order small. This allows us to retain only the first expansion term in the characteristic equation.

8 EQUATIONS OF MOTION OF A BODY–FLUID SYSTEM IN AN EQUIVALENT FORM

To solve the problem of optimal control of rotation for a body–fluid system, it is convenient to have an expression for the perturbation of the angular velocity Ω as a function of the control moment M.

In the further analysis, we leave only one term in the infinite sum in the first equation of system (6.3).

We denote the Laplace transform of an original $S(t)$ by $\hat{S}(p)$; i.e.,

$$\hat{S}(p) = L[S(t)] = \int_0^\infty S(t)e^{-pt}dt.$$

Applying the Laplace transform to the equations of system (6.3), we express $\hat{S}_1 = L[S_1]$ from the second and the third equation and, substituting the expression into the first equation, obtain the following expression for $\hat{\Omega}(p)$ in the Laplace space:

$$\begin{aligned} \hat{\Omega}(p) &= \hat{M}(p)\hat{F}(p), \\ \hat{F}(p) &= \frac{(p - i\gamma_1)(p + i\gamma_1\delta_1)}{z_3(p^3 + iz_2p^2 - z_1p - iz_0)}. \end{aligned} \tag{8.1}$$

To determine the inverse Laplace transform, we use the convolution theorem and expansion theorems of operational calculus [52]. We have

$$\Omega(t) = \int_0^t M(\tau)F(t - \tau)d\tau. \tag{8.2}$$

Here

$$F(t) = \sum_{n=1}^{3} \frac{(p_n - i\gamma_1)(p_n + i\gamma_1\delta_1)}{z_3(3p_n^2 + 2iz_2p_n - z_1)} e^{p_nt}, \tag{8.3}$$

where $p_n = iq_n$ $(n = 1, 2, 3)$ are the roots of the denominator in expression (8.1).

The expression (8.3) for the function F can be rewritten in the form

$$F(t) = \sum_{n=1}^{3} iZ_n e^{iq_nt}, \tag{8.4}$$

where the Z_n and the q_n $(n = 1, 2, 3)$ are real parameters.

Recall that $\Omega = \Omega_1 - i\Omega_2$ and $M = M_1 - iM_2$. We introduce the notation

$$
\begin{aligned}
P_k(t) &= -Z_k \int_0^t [M_1(\tau)\sin q_k(t-\tau) - M_2(\tau)\cos q_k(t-\tau)]\,d\tau,\\
Q_k(t) &= -Z_k \int_0^t [M_1(\tau)\cos q_k(t-\tau) + M_2(\tau)\sin q_k(t-\tau)]\,d\tau
\end{aligned}
\tag{8.5}
$$

$(k = 1, 2, 3)$.

Taking into account relations (8.2)–(8.5), we obtain

$$
\Omega_1 = \sum_{k=1}^{3} P_k, \quad \Omega_2 = \sum_{k=1}^{3} Q_k. \tag{8.6}
$$

Thus, expression (8.2) is equivalent to a system of ordinary differential equations; introducing the notation

$$
x = (\Omega_1, \Omega_2, P_1, P_2, P_3, Q_1, Q_2, Q_3)
$$

for the phase vector and the constant matrices

$$
A_{n\times n} = \begin{pmatrix}
0 & 0 & 0 & 0 & 0 & q_1 & q_2 & q_3\\
0 & 0 & -q_1 & -q_2 & -q_3 & 0 & 0 & 0\\
0 & 0 & 0 & 0 & 0 & q_1 & 0 & 0\\
0 & 0 & 0 & 0 & 0 & 0 & q_2 & 0\\
0 & 0 & 0 & 0 & 0 & 0 & 0 & q_3\\
0 & 0 & -q_1 & 0 & 0 & 0 & 0 & 0\\
0 & 0 & 0 & -q_2 & 0 & 0 & 0 & 0\\
0 & 0 & 0 & 0 & -q_3 & 0 & 0 & 0
\end{pmatrix}
$$

and

$$
B_{n\times m} = \begin{pmatrix}
0 & -\sum_{k=1}^{3} Z_k & 0 & 0 & 0 & -Z_1 & -Z_2 & -Z_3\\
\sum_{k=1}^{3} Z_k & 0 & Z_1 & Z_2 & Z_3 & 0 & 0 & 0
\end{pmatrix}^T,
$$

where $n = 8$ and $m = 2$, we write this system, together with the initial conditions, in the form

$$
\begin{cases}
\dot{x}(t) = Ax(t) + BM(t),\\
x(0) = x_0.
\end{cases}
\tag{8.7}
$$

Chapter 3

Oscillations of a plate in a viscous fluid: The flat wall model

This chapter has a special place in the book. Here we state and solve an initial boundary value problem describing an unsteady flow of a viscous fluid in the half-space above a rotating flat wall. A fluid rotates bodily with the wall bounding it at angular velocity $\omega_0 = \text{const}$ about a direction not perpendicular to the wall. The unsteady flow is induced by the longitudinal oscillations of the wall. In this setting, the velocity field of the viscous flow tangent to the flat wall is found. The tangential stress vector acting from the fluid on the plate is also calculated. The solution is given in analytic form.

In the fourth Chapter, we shall consider the problem of the perturbed motion of a rotating rigid body with a cavity containing a viscous fluid. The shape of the cavity is arbitrary, and the perturbation is any function of time. The hydrodynamic part of the problem is solved by the boundary layer method. For the initial approximation is used the solution of the problem on the oscillations of a body with a cavity filled with an ideal fluid. The velocity field of a viscous fluid in the boundary layer is determined. It turns out that the tangential component of this velocity field coincides with the velocity field of the fluid in the problem of the oscillations of a flat wall in a half-space filled with a viscous incompressible fluid, provided that the velocity of the plate is replaced by the negative velocity of the ideal fluid. This result makes it possible to use the flat wall model for taking into account the influence of a viscous fluid on the motion of a rigid body containing this fluid. It is this observation which determines the important role of this chapter in further exposition.

The first section of this chapter studies a unsteady boundary layer on a rotating plate. We formulate an initial boundary value problem and give an analytical solution of the hydrodynamic equations describing the evolution of flows of a viscous incompressible fluid. The unsteady boundary layer induces a plane wall rotating at a constant angular velocity in a semi-infinite space filled with a viscous fluid. At the initial moment of time, the wall begins to perform longitudinal oscillations.

In Section 2 we consider longitudinal quasi-harmonic oscillations of a plate and introduce a complex structure on the velocity field. It is shown that the velocity vector of a viscous fluid and the vector of tangential stresses acting from the fluid on the plate can be expressed by means of special functions. Asymptotic expressions for the velocity field and the friction forces are given.

Section 3 studies the structure of boundary layers adjacent to the plate. It is established that the obtained solution is the superposition of two waves propagating along the Oy axis toward each other and decaying exponentially at distances equal to the thicknesses of the corresponding boundary layers. The resonance case (where the frequency of wall oscillations equals twice the frequency of rotation of the plate–fluid system) is given special attention.

In Section 4 the vector of tangential stresses acting from the fluid on the plate is determined by direct calculation. It is shown that, in the absence of rotations, the result coincides with that obtained in [296].

Section 5 is devoted to oscillatory solutions. In this case, the Navier–Stokes equations, which describe the evolution of the flow of the viscous fluid surrounding the plate, reduces to a system of ordinary differential equations. The solution of the system satisfies all boundary conditions, provided that there is no resonance, and the boundary layer on the plate has the structure of a Stokes layer. However, in the resonance case (in which the frequency of wall oscillations equals twice the rotation frequency of the plate–fluid system), the velocity field is of oscillatory character and does not tend to zero at infinity (this phenomenon is known as the hydrodynamic paradox).

Section 6 studies the oscillations of a viscous incompressible fluid above a porous wall in the presence of a medium injection (suction), which is modeled by a constant velocity normal to the plate. An initial boundary value problem is formulated. The fluid and the plate rotate as a solid body at a constant angular velocity. The unsteady flow is induced by longitudinal oscillations of the plate and by a medium injection through the porous surface of the wall. The structure of the velocity field and the boundary layers adjacent to the plate is determined.

Section 7 considers the case of the motion of a plate with a constant acceleration in the absence of medium injection, which is important for applications.

1 AN UNSTEADY BOUNDARY LAYER ON A ROTATING PLATE

In this section we give an exact solution of the initial boundary value problem for the Navier–Stokes equations describing the flow of a fluid induced by a flat plate rotating at constant angular velocity ω_0 and moving progressively in its plane with velocity $\boldsymbol{u}(t)$. We also calculate the vector of tangential stresses acting on the plate from the side of the fluid. In the absence of rotation, the solution coincides with the well-known solution of the problem on the unsteady motion of a fluid bounded by a moving flat wall [296]. We study the quasi-harmonic oscillations of the plate and a motion with a constant acceleration. In the special case of harmonic oscillations under the assumption that the rotation axis is perpendicular to the plane of the plate, the results coincide with those obtained in [309]. In of the cases mentioned above, conclusions about the asymptotic behavior of solutions as $t \to \infty$ are made.

Suppose that an infinite plate H rotates, together with a fluid, in space at an angular velocity $\omega_0 = \text{const}$ and moves in its own plane at a velocity $\boldsymbol{u}(t)$. The plate bounds a half-space Q filled with an incompressible fluid of density ρ and kinematic viscosity ν. The fluid is in the field of mass forces with potential U.

To the plate we attach a Cartesian coordinate system $Oxyz$ with basis vectors $\boldsymbol{e}_x$, $\boldsymbol{e}_y$, and $\boldsymbol{e}_z$ so that the plane Oxz coincides with the plane of the plate and the Oy axis is inside the fluid perpendicularly to the plate.

The equations of motion of the fluid in the $Oxyz$ system and the boundary and initial condition have the form

$$\begin{cases} \omega_0 \times (\omega_0 \times r) + 2\omega_0 \times V + \dfrac{\partial V}{\partial t} + (V\nabla)V = -\dfrac{\nabla P}{\rho} + \nabla U + \nu \Delta V, \\ \operatorname{div} V = 0, \quad r \in Q, \\ V(r,t) = u(t), \quad r \in H, \quad t > 0, \\ |V(r,t)| \to 0, \quad |r| \to \infty, \quad t > 0, \\ V(r,0) = 0, \quad r \in Q. \end{cases} \tag{1.1}$$

Here t is time, r is the radius vector from the origin O, V is the velocity of the fluid, and P is the pressure.

We seek a solution of equations (1.1) in the form

$$\begin{cases} P = \dfrac{\rho}{2}(\omega_0 \times r)^2 + \rho U + \rho q(y,t), \\ V = V_x(y,t)e_x + V_z(y,t)e_z. \end{cases} \tag{1.2}$$

In this case, system (1.1) decomposes into the two subsystems

$$\begin{cases} \dfrac{\partial V}{\partial t} + 2\Omega(e_y \times V) = \nu \dfrac{\partial^2 V}{\partial y^2}, \quad V = u(t), \quad y = 0, \quad t > 0, \\ |V| \to 0 \quad y \to \infty, \quad t > 0, \quad V(y,0) = 0, \quad y > 0, \end{cases} \tag{1.3}$$

where $\Omega = \omega_0 \cdot e_y$, and

$$\begin{cases} \dfrac{\partial q}{\partial y} = 2V \cdot (\omega_0 \times e_y), \\ q(y,t) \to 0 \quad \text{as } y \to \infty, \ t > 0. \end{cases} \tag{1.4}$$

The velocity field is found from equations (1.3), and the pressure field is determined from the found velocity field by equations (1.4).

We seek a solution of system (1.3) in the form

$$\begin{cases} V(y,t) = W(y,t)\sin 2\Omega t - W(y,t) \times e_y \cos 2\Omega t, \\ W(y,t) = W_x(y,t)e_x + W_z(y,t)e_z, \end{cases} \tag{1.5}$$

where $W(y,t)$ is an unknown function.

Substituting (1.5) into (1.3), we obtain the following problem for determining W:

$$\begin{cases} \dfrac{\partial W}{\partial t} = \nu \dfrac{\partial^2 W}{\partial y^2}, \\ W(0,t) = u(t)\sin 2\Omega t + u(t) \times e_y \cos 2\Omega t, \quad t > 0, \\ |W(y,t)| \to 0, \quad y \to \infty, \quad t > 0, \quad W(y,0) = 0, \quad y > 0. \end{cases} \tag{1.6}$$

The solution of equations (1.6) is given by the well-known expression [80]

$$\boldsymbol{W}(y,t) = \frac{y}{2\sqrt{\pi\nu}} \int_0^t \frac{\boldsymbol{W}(0,\tau)}{(t-\tau)^{3/2}} \exp\left[-\frac{y^2}{4\nu(t-\tau)}\right] d\tau. \tag{1.7}$$

Substituting (1.7) into (1.5), we obtain the velocity field in the form

$$\boldsymbol{V}(y,t) = \frac{y}{2\sqrt{\pi\nu}} \int_0^t \frac{\boldsymbol{T}(\tau,t)}{(t-\tau)^{3/2}} \exp\left[-\frac{y^2}{4\nu(t-\tau)}\right] d\tau, \tag{1.8}$$

where

$$\boldsymbol{T}(\tau,t) = \boldsymbol{u}(\tau)\cos 2\Omega(t-\tau) + \boldsymbol{u}(\tau) \times \boldsymbol{e}_y \sin 2\Omega(t-\tau). \tag{1.9}$$

Taking into account (1.8) and (1.9) on the right-hand side of (1.4) and solving the resulting equations, we find the pressure field:

$$q(y,t) = 2\sqrt{\frac{\nu}{\pi}}(\boldsymbol{\omega}_0 \times \boldsymbol{e}_y) \int_0^t \frac{\boldsymbol{T}(\tau,t)}{(t-\tau)^{1/2}} \exp\left[-\frac{y^2}{4\nu(t-\tau)}\right] d\tau. \tag{1.10}$$

The vector of tangential stresses acting on the plate from the side of the fluid is determined by the expression [305]

$$\boldsymbol{f} = \rho\nu \left.\frac{\partial \boldsymbol{V}}{\partial y}\right|_{y=0}. \tag{1.11}$$

Substituting the velocity $\boldsymbol{V}$ from (1.8) into (1.11) and performing simple but cumbersome calculations, we obtain

$$\boldsymbol{f} = -\rho\sqrt{\frac{\nu}{\pi}}\left[\int_0^t \frac{\frac{\partial \boldsymbol{T}(\tau,t)}{\partial \tau}}{(t-\tau)^{1/2}} d\tau + \frac{\boldsymbol{T}(0,t)}{\sqrt{t}}\right]. \tag{1.12}$$

Relations (1.8)–(1.10) and (1.12) completely solve the problem.

For further analysis, it is convenient to represent the velocity field and the tangential stress vector in complex form.

Let us introduce the complex velocity vectors of the fluid and the plate and the complex stress vector:

$$\hat{V} = V_z + iV_x, \quad \hat{u} = u_z + iu_x, \quad \hat{f} = f_z + if_x. \tag{1.13}$$

Using expressions (1.8)–(1.12), we obtain

$$\hat{V}(y,t) = \frac{y}{2\sqrt{\pi\nu}} \int_0^t \frac{\hat{u}(\tau)}{(t-\tau)^{3/2}} \exp\left[-i2\Omega(t-\tau) - \frac{y^2}{4\nu(t-\tau)}\right] d\tau, \tag{1.14}$$

$$\hat{f} = -\rho\sqrt{\frac{\nu}{\pi}} \int_0^t \frac{\frac{\partial}{\partial\tau}[\hat{u}(\tau)\exp(-i2\Omega(t-\tau))]}{(t-\tau)^{1/2}} d\tau. \tag{1.15}$$

2 LONGITUDINAL QUASI-HARMONIC OSCILLATIONS OF THE PLATE

Suppose that the plate oscillates in its plane at a speed of

$$\hat{u} = e^{-\alpha t}(\hat{A}_1 e^{i\omega t} + \hat{A}_2 e^{-i\omega t}), \tag{2.1}$$

where the $\hat{A}_j$ $(j = 1, 2)$ are complex constants, ω is the frequency, and α is the damping coefficient. In this case, the velocity field and the stress vector can be expressed in terms of special functions.

Substituting (2.1) into (1.14) and (1.15), we obtain

$$\hat{V} = \frac{y}{2\sqrt{\pi\nu}} e^{-i2\Omega t} \sum_{j=1}^{2} \hat{A}_j e^{p_j t} \int_0^t \exp\left(-p_j\theta - \frac{y^2}{4\nu\theta}\right) \frac{d\theta}{\theta^{3/2}}, \tag{2.2}$$

$$\hat{f} = -\rho\sqrt{\frac{\nu}{\pi}} e^{-i2\Omega t} \sum_{j=1}^{2} \hat{A}_j p_j e^{p_j t} \int_0^t \exp(-p_j\theta) \frac{d\theta}{\theta^{1/2}}, \tag{2.3}$$

where $\theta = t - \tau$, $p_1 = -\alpha + i(2\Omega + \omega)$, and $p_2 = -\alpha + i(2\Omega - \omega)$.

The change of variables $\xi = \sqrt{p\theta}$, $\mathrm{Re}\sqrt{p} > 0$, reduces the integral in expression (2.3), that is,

$$J_1 = \int_0^t \exp(-p\theta)\theta^{-1/2} d\theta, \tag{2.4}$$

to the form

$$J_1 = \sqrt{\frac{\pi}{p}} \,\mathrm{erf}(\sqrt{pt}), \quad \text{where } \mathrm{erf}\, x = \frac{2}{\sqrt{\pi}} \int_0^x e^{-\xi^2} d\xi. \tag{2.5}$$

The integral in (2.2) does not reduce to table integrals [80], and evaluating it requires a special trick:

$$J_2 = \frac{y}{2\sqrt{\pi\nu}} \int_0^t \exp\left(-p\theta - \frac{y^2}{4\nu\theta}\right) \frac{d\theta}{\theta^{3/2}}. \tag{2.6}$$

We write

$$p\theta + \frac{y^2}{4\nu\theta} = \left(\sqrt{p\theta} + \frac{y}{2\sqrt{\nu\theta}}\right)^2 - y\sqrt{\frac{p}{\nu}}, \quad \mathrm{Re}\sqrt{p} > 0. \tag{2.7}$$

The change of variables $\frac{y}{2\sqrt{\nu\theta}} = \zeta$ in (2.6) yields

$$J_2 = e^{y\sqrt{\frac{p}{\nu}}} \frac{2}{\sqrt{\pi}} \int_{\frac{y}{2\sqrt{\nu t}}}^{\infty} \exp\left[-\left(\sqrt{\frac{p}{\nu}}\frac{y}{2\zeta} + \zeta\right)^2\right] d\zeta. \tag{2.8}$$

Setting $\sqrt{\frac{p}{\nu}}\frac{y}{2\zeta} + \zeta = \xi$, we represent (2.8) in the form

$$J_2 = e^{y\sqrt{\frac{p}{\nu}}} \frac{2}{\sqrt{\pi}} \int_{\frac{y}{2\sqrt{\nu t}}+\sqrt{pt}}^{\infty} e^{-\xi^2} d\xi + \frac{y}{\sqrt{\pi}}\sqrt{\frac{p}{\nu}}\, e^{y\sqrt{\frac{p}{\nu}}} \int_{\frac{y}{2\sqrt{\nu t}}}^{\infty} e^{-\left(\zeta + \sqrt{\frac{p}{\nu}}\frac{y}{2\zeta}\right)^2} \frac{d\zeta}{\zeta^2}. \tag{2.9}$$

Evaluating (2.6) by a different scheme, we obtain

$$p\theta + \frac{y^2}{4\nu\theta} = \left(\frac{y}{2\sqrt{\nu\theta}} - \sqrt{p\theta}\right)^2 + y\sqrt{\frac{p}{\nu}}, \quad \mathrm{Re}\sqrt{p} > 0. \tag{2.10}$$

The change $\frac{y}{2\sqrt{\nu\theta}} = \zeta$ reduces (2.6) to the form

$$J_2 = e^{-y\sqrt{\frac{p}{\nu}}} \frac{2}{\sqrt{\pi}} \int_{\frac{y}{2\sqrt{\nu t}}}^{\infty} e^{-\left(\zeta - \frac{y}{2\zeta}\sqrt{\frac{p}{\nu}}\right)^2} d\zeta. \tag{2.11}$$

Setting $\xi = \zeta - \frac{y}{2\zeta}\sqrt{\frac{p}{\nu}}$ in (2.11), we obtain

$$J_2 = e^{-y\sqrt{\frac{p}{\nu}}} \frac{2}{\sqrt{\pi}} \int_{\frac{y}{2\sqrt{\nu t}}-\sqrt{pt}}^{\infty} e^{-\xi^2} d\xi - e^{-y\sqrt{\frac{p}{\nu}}} \frac{y}{\sqrt{\pi}}\sqrt{\frac{p}{\nu}} \int_{\frac{y}{2\sqrt{\nu t}}}^{\infty} e^{-\left(\zeta - \frac{y}{2\zeta}\sqrt{\frac{p}{\nu}}\right)^2} \frac{d\zeta}{\zeta^2}. \tag{2.12}$$

Summing (2.9) and (2.12), we obtain the following expression for (2.6) in terms of special functions:

$$J_2 = \frac{1}{2}e^{-y\sqrt{\frac{p}{\nu}}}\operatorname{erfc}\left(\frac{y}{2\sqrt{\nu t}} - \sqrt{pt}\right) + \frac{1}{2}e^{y\sqrt{\frac{p}{\nu}}}\operatorname{erfc}\left(\frac{y}{2\sqrt{\nu t}} + \sqrt{pt}\right), \tag{2.13}$$

where $\operatorname{erfc} x = 1 - \operatorname{erf} x$.

Taking into account (2.2) and (2.13), we can write the expressions for the velocity and tangential stress vector in the form

$$\hat{V} = \frac{e^{-i2\Omega t}}{2}\sum_{j=1}^{2}\hat{A}_j e^{p_j t}\left[e^{y\sqrt{\frac{p_j}{\nu}}}\operatorname{erfc}\left(\frac{y}{2\sqrt{\nu t}} + \sqrt{p_j t}\right) + e^{-y\sqrt{\frac{p_j}{\nu}}}\operatorname{erfc}\left(\frac{y}{2\sqrt{\nu t}} - \sqrt{p_j t}\right)\right], \tag{2.14}$$

$$\hat{f} = -\rho\sqrt{\nu}\left[e^{-2i\Omega t}\sum_{j=1}^{2}\hat{A}_j\sqrt{p_j}e^{p_j t}\operatorname{erf}\sqrt{p_j t} + \frac{\hat{u}(0)e^{-2i\Omega t}}{\sqrt{\pi t}}\right]. \tag{2.15}$$

Of interest is the behavior of the solution as $t \to \infty$. First, observe that

$$\begin{cases} \operatorname{erfc}\left(\frac{y}{2\sqrt{\nu t}} + \sqrt{p_j t}\right) \to 0 \quad \text{as } t \to \infty, \ \operatorname{Re}\sqrt{p_j} > 0, \\ \operatorname{erfc}\left(\frac{y}{2\sqrt{\nu t}} - \sqrt{p_j t}\right) \to 2 \quad \text{as } t \to \infty, \ \operatorname{Re}\sqrt{p_j} > 0. \end{cases} \tag{2.16}$$

Asymptotic expressions for the velocity field and the friction forces are

$$\hat{V} = e^{-\alpha t}\left(\hat{A}_1 e^{i\omega t - y\sqrt{\frac{p_1}{\nu}}} + \hat{A}_2 e^{-i\omega t - y\sqrt{\frac{p_2}{\nu}}}\right), \quad \alpha \neq 0, \tag{2.17}$$

$$\hat{f} = -\rho\sqrt{\nu}e^{-\alpha t}(\hat{A}_1\sqrt{p_1}e^{i\omega t} + \hat{A}_2\sqrt{p_2}e^{-i\omega t}). \tag{2.18}$$

3 BOUNDARY LAYER STRUCTURE

Let us study expression (2.17) in more detail. For convenience, we introduce the notation

$$\begin{aligned} k_{1,2} &= \frac{2\Omega \pm \omega}{\sqrt{2\nu}}[\alpha^2 + (2\Omega \pm \omega)^2]^{-\frac{1}{4}}, \\ \delta_{1,2} &= \left(\frac{2\nu}{2\alpha^2 + (2\Omega \pm \omega)^2}\right)^{\frac{1}{2}}[\alpha^2 + (2\Omega \pm \omega)^2]^{\frac{1}{4}}. \end{aligned} \tag{3.1}$$

Here the plus sign in (3.1) corresponds to the index 1 and the minus sign, to the index 2. Taking into account (3.1), we can write the velocity field (2.17) in the form

$$\hat{V} = e^{-\alpha t}\left(\hat{A}_1 e^{-\frac{y}{\delta_1}} e^{i(\omega t - k_1 y)} + \hat{A}_2 e^{-\frac{y}{\delta_2}} e^{-i(\omega t - k_2 y)}\right). \tag{3.2}$$

This solution is the superposition of two waves with wave numbers k_j ($j = 1, 2$) and frequency ω, which propagate along the Oy axis toward each other and exponentially decay at distances of respective orders δ_j.

Solution (3.2) is feasible uniformly over the entire domain both in the nonresonant and the resonant case ($2\Omega = \omega$). Indeed, at $2\Omega = \omega$, we have

$$k_1 = \frac{4\Omega}{\sqrt{2\nu}}\left(\alpha^2 + 16\Omega^2\right)^{-\frac{1}{4}}, \quad \delta_1 = \frac{4\Omega}{k_1}\left(2\alpha^2 + 16\Omega^2\right)^{-\frac{1}{2}},$$
$$k_2 = 0, \quad \delta_2 = \sqrt{\frac{\nu}{\alpha}}; \tag{3.3}$$

thus, in the resonance case, there is no wave incident to the plate, but the solution still decays into the fluid. At $\alpha = 0$, however, $\delta_2 \to \infty$, and the solution becomes unfeasible as $y \to \infty$, because the thickness of one of the boundary layers unboundedly increases. This effect of the absence of an oscillatory solution at $2\Omega = \omega$ was discussed in [309].

This leads us to the conclusion that damping removes the difficulties mentioned in [309]. In this sense, its role is similar to that played by the suction of a fluid from the surface of a porous plate, which was considered in [89].

As follows from (2.14), (2.16) and (3.1), (3.2), at $2\Omega = \omega$ and $\alpha = 0$, the stable solution is uniformly feasible in the entire region occupied by the fluid, has the form

$$\hat{V} = \hat{A}_1 e^{i\omega t} e^{-y\sqrt{\frac{p_1}{\nu}}} + \hat{A}_2 e^{-i\omega t}\,\mathrm{erfc}\left(\frac{y}{2\sqrt{\nu t}}\right). \tag{3.4}$$

Solution (3.4) is of oscillatory character. The distance from the plate on which this difference becomes substantial is $y = o(\sqrt{\nu t})$. When the angular rotation velocity $\boldsymbol{\omega}_0$ of the plate is perpendicular to the plate plane, (3.4) transforms into the solution obtained earlier for this case in [309].

Let us study the exponentially decaying motion of the plate. In relations (2.14) and (2.15), we set

$$\omega = 0, \quad \hat{A}_1 + \hat{A}_2 = \hat{A}, \quad \sigma = 2i\Omega - \alpha; \tag{3.5}$$

then the velocity field and the tangential stress vector take the form

$$\hat{V} = \frac{1}{2}\hat{A}e^{-\alpha t}\left[e^{y\sqrt{\frac{\sigma}{\nu}}}\,\mathrm{erfc}\left(\frac{y}{2\sqrt{\nu t}} + \sqrt{\sigma t}\right) + e^{-y\sqrt{\frac{\sigma}{\nu}}}\,\mathrm{erfc}\left(\frac{y}{2\sqrt{\nu t}} - \sqrt{\sigma t}\right)\right],$$
$$\hat{f} = -\rho\sqrt{\sigma\nu}\hat{A}\,\mathrm{erf}(\sqrt{\sigma t})e^{-\alpha t}, \quad \mathrm{Re}\sqrt{\sigma} > 0. \tag{3.6}$$

Later on, we shall consider a steady solutions and its characteristics. Using the asymptotic representation of the additional error function as $t \to \infty$, we obtain

$$\hat{V} = \hat{A}e^{-y\sqrt{\frac{\alpha}{\nu}}-\alpha t}, \quad \hat{f} = -\rho\sqrt{\sigma\nu}\hat{A}e^{-\alpha t}. \tag{3.7}$$

In a similar way, on the basis of the general solution (1.8), other cases of the nonstationary motion of the plate can be analyzed.

Thus, if we seek a solution in the form (1.2), then the nonlinear system of Navier–Stokes nonstationary 3D equations decomposes into two subsystems, one of which determines the velocity field and the other, the pressure field. The substitution of (1.5) removes the "curl" term from the equation and reduces the leads system to a heat-type equation, which makes it possible to obtain an exact solution for the velocity field and the pressure.

For case of the motion of the plate in its plane at velocity (2.1), the velocity field can be expressed in terms of special functions.

As $t \to \infty$, the velocity field is the superposition of two waves propagating along the Oy axis toward each other and exponentially decaying at respective distances δ_j.

4 TANGENTIAL STRESS VECTOR

In this section we directly calculate the vector of tangential stresses acting from the fluid on the plate. As is known,

$$f = \rho\nu \frac{\partial V}{\partial y}\bigg|_{y=0}.$$

In the first section of this chapter, we found the velocity field of the fluid induced by the motion of the plate at a velocity $\boldsymbol{u}(t)$. Consider this expression:

$$V(y,t) = \frac{y}{2\sqrt{\pi\nu}} \int_0^t \frac{T(\tau,t)}{(t-\tau)^{3/2}} \exp\left[-\frac{y^2}{4\nu(t-\tau)}\right] d\tau, \tag{4.1}$$

where

$$T(\tau,t) = \boldsymbol{u}(\tau)\cos 2\Omega(t-\tau) + \boldsymbol{u}(\tau) \times \boldsymbol{e}_y \sin 2\Omega(t-\tau).$$

Let us make the change of variables

$$\frac{y^2}{4\nu(t-\tau)} = z, \quad t-\tau = \frac{y^2}{4\nu z}, \quad \tau = t - \frac{y^2}{4\nu z}, \quad d\tau = \frac{y^2}{4\nu z^2}dz.$$

The limits of integration change. Thus, the upper limit at $\tau = t$ becomes $z = \infty$, and the lower limit at $\tau = 0$ becomes $z = \frac{y^2}{4\nu t}$. We have

$$V(y,t) = \frac{1}{\sqrt{\pi}} \int_{\frac{y^2}{4\nu t}}^{\infty} \left[\boldsymbol{u}\left(t - \frac{y^2}{4\nu z}\right)\cos 2\Omega\frac{y^2}{4\nu z} + \boldsymbol{u}\left(t - \frac{y^2}{4\nu z}\right) \times \boldsymbol{e}_y \sin 2\Omega\frac{y^2}{4\nu z}\right] e^{-z}\frac{dz}{\sqrt{z}}. \tag{4.2}$$

Applying the formula for differentiation under the integral sign in the case where the limits of integration depend on the parameter with respect to which the differentiation is performed, we obtain

$$\sqrt{\pi}\frac{\partial V(y,t)}{\partial y} = -\frac{1}{\sqrt{\nu}}e^{-\frac{y^2}{4\nu t}}\left(\frac{\boldsymbol{u}(0)\cos 2\Omega t + \boldsymbol{u}(0)\times \boldsymbol{e}_y \sin 2\Omega t}{\sqrt{t}}\right)$$
$$+\int_{\frac{y^2}{4\nu t}}^{\infty}\frac{\partial}{\partial y}\left[\boldsymbol{u}\left(t-\frac{y^2}{4\nu z}\right)\cos 2\Omega\frac{y^2}{4\nu z} + \boldsymbol{u}\left(t-\frac{y^2}{4\nu z}\right)\times \boldsymbol{e}_y \sin 2\Omega\frac{y^2}{4\nu z}\right]e^{-z}\frac{dz}{\sqrt{z}}.$$

Consider separately the derivative of the integrand:

$$\frac{\partial}{\partial y}\left[\boldsymbol{u}\left(t-\frac{y^2}{4\nu z}\right)\cos 2\Omega\frac{y^2}{4\nu z} + \boldsymbol{u}\left(t-\frac{y^2}{4\nu z}\right)\times \boldsymbol{e}_y \sin 2\Omega\frac{y^2}{4\nu z}\right]$$
$$= -\frac{2y}{4\nu z}\frac{\partial}{\partial \tau}\left[\boldsymbol{u}\left(t-\frac{y^2}{4\nu z}\right)\cos 2\Omega\frac{y^2}{4\nu z} + \boldsymbol{u}\left(t-\frac{y^2}{4\nu z}\right)\times \boldsymbol{e}_y \sin 2\Omega\frac{y^2}{4\nu z}\right]$$

We return to the initial variables:

$$\sqrt{\pi}\frac{\partial V}{\partial y} = -\frac{1}{\sqrt{\nu}}e^{-\frac{y^2}{4\nu t}}\left[\frac{\boldsymbol{u}(0)\cos 2\Omega t + \boldsymbol{u}(0)\times \boldsymbol{e}_y \sin 2\Omega t}{\sqrt{t}}\right]$$
$$-\int_0^t \frac{\partial}{\partial \tau}[\boldsymbol{u}(\tau)\cos 2\Omega(t-\tau) + \boldsymbol{u}(\tau)\times \boldsymbol{e}_y \sin 2\Omega(t-\tau)]\frac{e^{-\frac{y^2}{4\nu(t-\tau)}}2y4\nu z^2}{4\nu z\sqrt{z}y^2}d\tau.$$

Finally, we have

$$\sqrt{\pi}\frac{\partial V}{\partial y} = -\frac{\boldsymbol{u}(0)\cos 2\Omega t + \boldsymbol{u}(0)\times \boldsymbol{e}_y \sin 2\Omega t}{\sqrt{\nu t}}e^{-\frac{y^2}{4\nu t}}$$
$$-\frac{1}{\sqrt{\nu}}\int_0^t \frac{\frac{\partial}{\partial \tau}[\boldsymbol{u}(\tau)\cos 2\Omega(t-\tau) + \boldsymbol{u}(\tau)\times \boldsymbol{e}_y \sin 2\Omega(t-\tau)]}{\sqrt{t-\tau}}e^{-\frac{y^2}{4\nu(t-\tau)}}d\tau. \quad (4.3)$$

Now, it is easy to find the vector $\boldsymbol{f}$:

$$\boldsymbol{f} = -\rho\sqrt{\frac{\nu}{\pi}}\left[\frac{\boldsymbol{u}(0)\cos 2\Omega t + \boldsymbol{u}(0)\times \boldsymbol{e}_y \sin 2\Omega t}{\sqrt{t}}\right.$$
$$\left.+\int_0^t \frac{\frac{\partial}{\partial \tau}[\boldsymbol{u}(\tau)\cos 2\Omega(t-\tau) + \boldsymbol{u}(\tau)\times \boldsymbol{e}_y \sin 2\Omega(t-\tau)]}{\sqrt{t-\tau}}d\tau\right]. \quad (4.4)$$

Setting $\omega_0 = 0$ in (4.4), we arrive at the following well-known expression for the vector of tangential stresses acting from a fluid on a plate moving in its plane at a velocity $\boldsymbol{u}(t)$ [296]:

$$f = -\rho\sqrt{\frac{\nu}{\pi}}\left[\frac{\boldsymbol{u}(0)}{\sqrt{t}} + \int_0^t \frac{\boldsymbol{u}'(\tau)}{\sqrt{t-\tau}}d\tau\right]. \tag{4.5}$$

Expressions (4.4) and (4.5) serve as a basis calculating the dissipative forces acting on a body performing rotational and librational motions from the side of a fluid filling a cavity in the body.

5 OSCILLATORY SOLUTIONS

Suppose that the bounding plane H rotates in space together with the fluid at an angular velocity $\omega_0 =$ const and suddenly begins to move in the longitudinal direction at a rate $\boldsymbol{u} = \boldsymbol{u}(y)e^{\lambda t}$, where $\boldsymbol{u}(y) = \{u_x(y), 0, u_z(y)\}$.

For the quasi-harmonic mode, we have

$$\boldsymbol{v}(y,t) = \boldsymbol{v}(y)e^{\lambda t}, \quad \boldsymbol{v}(y) = \{v_x(y), 0, v_z(y)\}.$$

In this case, the Navier–Stokes equations and the boundary conditions can be written in the form

$$\lambda v_x + 2(\omega_0 e_y)v_z = \nu\frac{d^2 v_x}{dy^2},$$
$$\lambda v_z - 2(\omega_0 e_y)v_x = \nu\frac{d^2 v_z}{dy^2}, \tag{5.1}$$
$$v_x(0) = u_x(0), \quad v_z(0) = u_z(0), \quad v_{x,z} \to 0 \quad \text{as } y \to \infty.$$

This is a system of equations with constant coefficients for the functions v_x and v_z. The role of the argument is played by y.

The characteristic equation of the system (5.1) is

$$\mu^4 - 2\frac{\lambda}{\nu}\mu^2 + \left(\frac{\lambda}{\nu}\right)^2 + \left(\frac{2\omega_0 e_y}{\nu}\right)^2 = 0. \tag{5.2}$$

This equation has roots

$$\mu_{1,2} = \sqrt{\frac{\lambda \pm i2\omega_0 e_y}{\nu}};$$

we have chosen those branches for which $\operatorname{Re}\mu_j \leqslant 0, j = 1,2.$

The solution of system (5.1) exists at $\lambda \neq \pm 2i\omega_0 e_y$; under the given boundary conditions, it is given by the expressions

$$
\begin{aligned}
v_x(y) &= -\frac{1}{2}[iu_z(0) - u_x(0)]E_1 + \frac{1}{2}[iu_z(0) + u_x(0)]E_2,\\
v_z(y) &= \frac{1}{2}[u_z(0) + iu_x(0)]E_1 + \frac{1}{2}[u_z(0) - iu_x(0)]E_2,
\end{aligned}
\tag{5.3}
$$

where $E_j = \exp(\mu_j y), j = 1, 2, \boldsymbol{v}(y) = \boldsymbol{u}(0)\frac{E_1+E_2}{2} + i\boldsymbol{u}(0) \times \boldsymbol{e}_y\frac{E_1-E_2}{2}$.

Finally, we obtain the following velocity field for the quasi-harmonic mode:

$$
\boldsymbol{v}(y,t) = e^{\lambda t}\left[\boldsymbol{u}(0)\frac{E_1+E_2}{2} + i\boldsymbol{u}(0) \times \boldsymbol{e}_y\frac{E_1-E_2}{2}\right]. \tag{5.4}
$$

Solution (5.4) satisfies all boundary conditions at $\lambda \neq \pm i2\omega_0 e_y$, and the boundary layer on the plate has the same structure as the Stokes layer.

However, at $\lambda = \pm i2\omega_0 e_y$, solution (5.4) has the form

$$
\boldsymbol{v}(y,t) = e^{\lambda t}\left[\boldsymbol{u}(0)\frac{E_1+1}{2} + i\boldsymbol{u}(0) \times \boldsymbol{e}_y\frac{E_1-1}{2}\right] \quad \text{at } \lambda = +2\omega_0 \boldsymbol{e}_y,
$$

$$
\boldsymbol{v}(y,t) = e^{\lambda t}\left[\boldsymbol{u}(0)\frac{1+E_2}{2} + i\boldsymbol{u}(0) \times \boldsymbol{e}_y\frac{1-E_2}{2}\right] \quad \text{at } \lambda = -2\omega_0 \boldsymbol{e}_y.
$$

In both cases, as $y \to \infty$, the velocity field

$$
\boldsymbol{v}(y,t) = \frac{1}{2}e^{\lambda t}[\boldsymbol{u}(0) \mp i\boldsymbol{u}(0) \times \boldsymbol{e}_y]
$$

is of oscillatory character and does not tend to zero, although remains bounded. In this resonance case $\lambda = \pm i2\omega_0 e_y$, the solution satisfies the boundary conditions on the plate H but does not satisfy the conditions at infinity.

Interestingly, if there is another infinite plate H_1 parallel to the first one at a distance of $y = l$ from it, then the boundary conditions can be satisfied by setting

$$
\boldsymbol{v}(y,t) = \frac{1}{2}e^{\lambda t}\left\{[\boldsymbol{u}(0) + i\boldsymbol{u}(0) \times \boldsymbol{e}_y]\frac{\operatorname{sh}\mu(l-y)}{\operatorname{sh}\mu l} + [\boldsymbol{u}(0) - i\boldsymbol{u}(0) \times \boldsymbol{e}_y]\frac{l-y}{l}\right\}.
$$

Here the first term represents a modified Stokes layer and the second, Kutta oscillations. In the next chapter we discuss this situation in detail.

6 OSCILLATIONS OF A VISCOUS INCOMPRESSIBLE FLUID ABOVE A POROUS PLATE IN THE PRESENCE OF MEDIUM INJECTION (SUCTION)

In [89] a solution of Navier–Stokes 3D stationary equations was obtained for a fluid flowing around a plate in the presence of uniform suction in a rotating coordinate system. The structure of the stationary velocity field was studied and a boundary layer on the porous plate was formed. It was shown that, in the presence of suction, the thickness of the boundary layer decreases. If a medium is injected through the plate surface, then, for a steady flow around the plate, there exists no asymptotic solution in the inertial system, while in the rotating, such a solution do exists. Thereby, it was proved that rotation is fully responsible for the existence of a solution in the case of injection.

From the point of view of generalization of results obtained in [89], this book studies the unsteady boundary layer of a flow in a viscous incompressible homogeneous fluid surrounding an infinite porous plate in the presence of uniform suction or injection of a medium. The fluid and the plate rotate as a solid body at a constant angular velocity. The plate moves in its plane at a velocity $\boldsymbol{u}(t)$.

The unsteady flow is induced by nontorsional oscillations of the plate. We determine the structure of the unsteady velocity field and of the formed boundary layers adjacent to the plate. It turns out that, in this case (in the presence of medium injection or suction), an exact solution of the unsteady Navier–Stokes 3D equations can be found too.

Consider an infinite plate moving in a fluid at a velocity $\boldsymbol{u}(t)$ in its plane. Through the surface of the plate, a medium is injected or sucked at a speed $u_1(t)$ along the normal to the plate surface. The medium is a viscous incompressible fluid occupying the half-space bounded by the plate. The motion of the fluid is described by the Navier–Stokes equations and boundary conditions which have the following form in the usual notation:

$$\begin{cases} (\boldsymbol{V}\nabla)\boldsymbol{V} + \dfrac{\partial \boldsymbol{V}}{\partial t} + 2\boldsymbol{\omega}_0 \times \boldsymbol{V} = -\nabla p + \nu\Delta\boldsymbol{V}, \\ \operatorname{div}\boldsymbol{V} = 0 \quad \text{on } Q, \\ \boldsymbol{V} = \{\boldsymbol{u}(t), u_1(t)\boldsymbol{e}_y\} \quad \text{for } \boldsymbol{r} \in H,\ t > 0, \\ |\boldsymbol{V}| \to 0 \quad \text{as } |\boldsymbol{r}| \to \infty,\ t > 0; \end{cases} \tag{6.1}$$

here $\boldsymbol{e}_y$ is a basis vector of the Cartesian coordinate system $Oxyz$ attached to the plate. The Oxz plane coincides with the plane of the plate.

The motion of the fluid begins from rest, so that

$$\boldsymbol{V}(0,\boldsymbol{r}) = 0 \quad \text{for } \boldsymbol{r} \in Q. \tag{6.2}$$

We seek a solution of system (6.1) in the form

$$\begin{cases} \boldsymbol{V} = \{V_x(y,t), u_1(t), V_z(y,t)\}, \\ p = 2\omega_{0z}u_1(t)x - 2\omega_{0x}u_1(t)z - \dfrac{\partial u_1}{\partial t}y + q(t,y). \end{cases} \tag{6.3}$$

For determining the velocity field, we obtain the system of equations

$$\frac{\partial V_x}{\partial t} + 2\omega_{0y} V_z = LV_x,$$
$$\frac{\partial V_z}{\partial t} - 2\omega_{0y} V_x = LV_z, \tag{6.4}$$
$$\frac{\partial q}{\partial y} = 2(\omega_{0z} V_x - \omega_{0x} V_z),$$

where $L = \nu\frac{\partial^2}{\partial y^2} - u_1(t)\frac{\partial}{\partial y}$.

We seek a solution of system (6.4) in the form

$$\mathbf{V} = \mathbf{W} \sin 2\Omega t - (\mathbf{W} \times \mathbf{e}_y) \cos 2\Omega t, \tag{6.5}$$

where $\mathbf{W}$ is an unknown function and $\omega_{0y} = \Omega$.

The unknown function $\mathbf{W}(y,t)$ satisfies a parabolic differential equation and the boundary conditions

$$\begin{cases} \dfrac{\partial \mathbf{W}}{\partial t} = L\mathbf{W}, \\ \mathbf{W} = \mathbf{u}(t) \sin 2\Omega t + (\mathbf{u}(t) \times \mathbf{e}_y) \cos 2\Omega t, \quad \mathbf{r} \in H, \quad t > 0, \\ \mathbf{W} \to 0 \quad \text{for } t = 0, y > 0, \\ \mathbf{W} \to 0 \quad \text{as } y \to \infty, \end{cases} \tag{6.6}$$

First, consider the case where $u_1(t) = \text{const} = a$, which corresponds to uniform suction or injection; $a > 0$ corresponds to a medium injection through the surface of the porous plate, and $a < 0$ corresponds to a suction of the surrouning medium.

Using Duhamel integral, we can write the solution of problem (6.6) as

$$\mathbf{W}(y,t) = \frac{d}{dt}\int_0^t [\mathbf{u}(\tau) \sin 2\Omega\tau + (\mathbf{u}(\tau) \times \mathbf{e}_y) \cos 2\Omega\tau] W_1(y, t-\tau) d\tau, \tag{6.7}$$

where W_1 is a solution of the boundary value problem

$$\frac{\partial W_1}{\partial t} + a\frac{\partial W_1}{\partial y} = \nu\frac{\partial^2 W_1}{\partial y^2},$$
$$W_1(0,t) = \begin{cases} 1, & t > 0, \\ 0, & t < 0, \end{cases} \tag{6.8}$$
$$|W_1| \to 0 \quad \text{as } y \to \infty.$$

We solve system (6.8) by the Laplace transform method. We define the Laplace image of functions by

$$\hat{\mathbf{u}}(p) = \int_0^\infty e^{-pt} \mathbf{u}(t) dt. \tag{6.9}$$

In the space of images, the differential equation (6.8) has the form

$$\begin{cases} p\hat{W}_1 + a\dfrac{\partial \hat{W}_1}{\partial y} = \nu\dfrac{\partial^2 \hat{W}_1}{\partial y^2}, \\ \hat{W}_1\Big|_{y=0} = \dfrac{l}{p}, \\ \hat{W}_1 = 0 \quad \text{as } y \to \infty, \end{cases} \tag{6.10}$$

where l is a constant. The solution of system (6.10) is

$$\begin{cases} \hat{W}_1(p, y) = \dfrac{l}{p}e^{-\lambda y}, \quad \text{where } \operatorname{Re}\lambda > 0, \\ \lambda = \dfrac{a}{2\nu} + \sqrt{\dfrac{a^2}{4\nu^2} + \dfrac{p}{\nu}}. \end{cases} \tag{6.11}$$

Passing to originals, we write solution (6.11) in the form [80]

$$W_1(y,t) = \frac{1}{2}le^{-\frac{ay}{2\nu}}\left[e^{-\frac{ay}{2\nu}}\operatorname{erfc}\left(\frac{y}{2\sqrt{\nu t}} - \frac{a}{2}\sqrt{\frac{t}{\nu}}\right) + e^{\frac{ay}{2\nu}}\operatorname{erfc}\left(\frac{y}{2\sqrt{\nu t}} + \frac{a}{2}\sqrt{\frac{t}{\nu}}\right)\right]. \tag{6.12}$$

Thus, the solution of system (6.6) is given by formulas (6.7) and (6.12).

Substituting (6.7) and taking into account (6.12) in (6.5), we obtain the required velocity field of the initial problem:

$$\begin{aligned} \boldsymbol{V} = \sin 2\Omega t\Bigg[&\frac{d}{dt}\int_0^t (\boldsymbol{u}(\tau)\sin 2\Omega\tau + (\boldsymbol{u}(\tau)\times \boldsymbol{e}_y)\cos 2\Omega\tau) \\ &\times \frac{1}{2}\left(e^{-\frac{ay}{\nu}}\operatorname{erfc}\left(\frac{y}{2\sqrt{\nu(t-\tau)}} - \frac{a}{2}\sqrt{\frac{t-\tau}{\nu}}\right)\right. \\ &\left.+\operatorname{erfc}\left(\frac{y}{2\sqrt{\nu(t-\tau)}} + \frac{a}{2}\sqrt{\frac{t-\tau}{\nu}}\right)\right)d\tau\Bigg] \\ - \cos 2\Omega t\Bigg[&\boldsymbol{e}_y \times \frac{d}{dt}\int_0^t (\boldsymbol{u}(\tau)\sin 2\Omega\tau + (\boldsymbol{u}(\tau)\times \boldsymbol{e}_y)\cos 2\Omega\tau) \\ &\times \frac{1}{2}\left(e^{-\frac{ay}{\nu}}\operatorname{erfc}\left(\frac{y}{2\sqrt{\nu(t-\tau)}} - \frac{a}{2}\sqrt{\frac{t-\tau}{\nu}}\right)\right. \\ &\left.+\operatorname{erfc}\left(\frac{y}{2\sqrt{\nu(t-\tau)}} + \frac{a}{2}\sqrt{\frac{t-\tau}{\nu}}\right)\right)d\tau\Bigg]. \end{aligned} \tag{6.13}$$

7 MOTION OF THE PLATE WITH A CONSTANT ACCELERATION

Consider the case where the plate moves with a constant acceleration, which is important for applications.

It is convenient to perform further transformations in complex form. We introduce the complex vectors

$$\hat{V} = V_z + iV_x, \quad \hat{u} = u_z + iu_x. \tag{7.1}$$

Using (7.1) and setting $a = 0$ (which corresponds to the absence of injection), we rewrite the velocity field (1.13) as

$$\hat{V} = \frac{y}{2\sqrt{\pi\nu}} \int_0^t \frac{\hat{u}(\tau)}{(t-\tau)^{3/2}} e^{-i2\Omega(t-\tau) - \frac{y^2}{4\nu(t-\tau)}} d\tau. \tag{7.2}$$

Of interest is the determination of the viscous stresses acting from the side of the fluid on the plate:

$$\hat{\tau} = -\rho\sqrt{\frac{\nu}{\pi}} \int_0^t \frac{\frac{\partial}{\partial\tau}\left[\hat{u}(\tau)e^{-i2\Omega(t-\tau)}\right]}{\sqrt{t-\tau}} d\tau. \tag{7.3}$$

Setting $\hat{u} = \hat{a}t$, where $\hat{a}$ is a constant vector, we obtain

$$\hat{V} = \frac{y}{2\sqrt{\pi\nu}} \int_0^t \frac{\hat{u}(\tau)}{(t-\tau)^{3/2}} e^{-i2\Omega(t-\tau) - \frac{y^2}{4\nu(t-\tau)}} d\tau$$

$$= \frac{\hat{a}y}{2\sqrt{\pi\nu}} \int_0^t \frac{\tau e^{-i2\Omega(t-\tau) - \frac{y^2}{4\nu(t-\tau)}}}{(t-\tau)^{3/2}} d\tau.$$

Introducing the notation $i2\Omega = \sigma$ and making the change of variables $t - \tau = \theta$, $\tau = t - \theta$, we arrive at

$$\hat{V} = \frac{\hat{a}y}{2\sqrt{\pi\nu}} \int_0^t \frac{t-\theta}{\theta^{3/2}} e^{-\left(\sigma\theta + \frac{y^2}{4\nu\theta}\right)} d\theta. \tag{7.4}$$

The integral in (7.4) can be split into two parts as

$$\hat{V} = \frac{\hat{a}y}{2\sqrt{\pi\nu}} t \int_0^t \frac{e^{-\left(\sigma\theta + \frac{y^2}{4\nu\theta}\right)}}{\theta^{3/2}} d\theta - \frac{\hat{a}y}{2\sqrt{\pi\nu}} \int_0^t \frac{e^{-\left(\sigma\theta + \frac{y^2}{4\nu\theta}\right)}}{\theta^{1/2}} d\theta;$$

but we have already evaluated J_1 and found that

$$J_1 = \frac{\hat{a}t}{2}\left[e^{-y\sqrt{\frac{\sigma}{\nu}}}\operatorname{erfc}\left(\frac{y}{2\sqrt{\nu t}} - \sqrt{\sigma t}\right) + e^{y\sqrt{\frac{\sigma}{\nu}}}\operatorname{erfc}\left(\frac{y}{2\sqrt{\nu t}} + \sqrt{\sigma t}\right)\right].$$

Let us evaluate the integral J_2:

$$J_2 = \frac{\hat{a}y}{2\sqrt{\pi\nu}}\int_0^t \frac{e^{-\left(\sigma\theta+\frac{y^2}{4\nu\theta}\right)}}{\sqrt{\theta}}d\theta.$$

First, we evaluate J_2 by the scheme

$$\sigma\theta + \frac{y^2}{4\nu\theta} = \left(\sqrt{\sigma\theta} - \frac{y}{2\sqrt{\nu\theta}}\right)^2 - y\sqrt{\frac{\sigma}{\nu}},$$

$$\frac{y}{2\sqrt{\nu\theta}} = \zeta,$$

$$\frac{y}{2\sqrt{\nu}}\frac{d\theta}{\sqrt{\theta}} = -2\theta d\zeta = -\frac{y^2}{2\nu\xi^2}d\xi,$$

$$J_2 = \frac{\hat{a}y^2}{2\nu\sqrt{\pi}}e^{y\sqrt{\frac{\sigma}{\nu}}}\int_{\frac{y}{2\sqrt{\nu t}}}^{\infty} e^{-\left(\xi+\frac{y}{2\xi}\sqrt{\frac{\sigma}{\nu}}\right)^2}\frac{d\xi}{\zeta^2}$$

$$= -\frac{\hat{a}y}{\sqrt{\pi\sigma\nu}}e^{y\sqrt{\frac{\sigma}{\nu}}}\int_{\frac{y}{2\sqrt{\nu t}}}^{\infty} e^{-\left(\xi+\frac{y}{2\xi}\sqrt{\frac{\sigma}{\nu}}\right)^2}\left(-\frac{y}{2}\sqrt{\frac{\sigma}{\nu}}\frac{d\xi}{\xi^2}\right).$$

Setting

$$\zeta + \frac{y}{2\zeta}\sqrt{\frac{\sigma}{\nu}} = \xi, \quad \left(1 - \frac{y}{2\zeta^2}\sqrt{\frac{\sigma}{\nu}}\right)d\zeta = d\xi,$$

we obtain

$$J_2 = -\frac{\hat{a}y}{\sqrt{\pi\sigma\nu}}e^{y\sqrt{\frac{\sigma}{\nu}}}\int_{\frac{y}{2\sqrt{\nu t}}}^{\infty} e^{-\xi^2}d\xi - \frac{\hat{a}y}{\sqrt{\pi\sigma\nu}}e^{y\sqrt{\frac{\sigma}{\nu}}}\int_{\frac{y}{2\sqrt{\nu t}}}^{\infty} e^{-\left(\zeta+\frac{y}{2\zeta}\sqrt{\frac{\sigma}{\nu}}\right)^2}d\zeta,$$

or

$$J_2 = -\frac{\hat{a}y}{2\sqrt{\sigma\nu}}e^{y\sqrt{\frac{\sigma}{\nu}}}\operatorname{erfc}\left(\frac{y}{2\sqrt{\nu t}} + \sqrt{\sigma t}\right) + \frac{\hat{a}y}{\sqrt{\pi\sigma\nu}}e^{y\sqrt{\frac{\sigma}{\nu}}}\int_{\frac{y}{2\sqrt{\nu t}}}^{\infty} e^{-\left(\zeta+\frac{y}{2\zeta}\sqrt{\frac{\sigma}{\nu}}\right)^2}d\zeta. \quad (7.5)$$

Let us evaluate J_2 in a different way:

$$\sigma\theta + \frac{y^2}{4\nu\theta} = \left(\frac{y}{2\sqrt{\nu\theta}} - \sqrt{\sigma\theta}\right)^2 + y\sqrt{\frac{\sigma}{\nu}},$$

$$\frac{y}{2\sqrt{\nu\theta}} = \zeta, \quad \frac{y}{2\sqrt{\nu}}\frac{d\theta}{\sqrt{\theta}} = -2\theta d\zeta = -\frac{y^2}{2\nu\zeta^2}d\zeta,$$

$$J_2 = \frac{\hat{a}y^2}{2\nu\sqrt{\pi}}e^{-y\sqrt{\frac{\sigma}{\nu}}}\int_{\frac{y}{2\sqrt{\nu t}}}^{\infty} e^{-\left(\zeta - \frac{y}{2\zeta}\sqrt{\frac{\sigma}{\nu}}\right)^2}\frac{d\zeta}{\zeta^2}.$$

We set

$$\zeta - \frac{y}{2\zeta}\sqrt{\frac{\sigma}{\nu}} = \xi, \quad \left(1 + \frac{y}{2\zeta^2}\sqrt{\frac{\sigma}{\nu}}\right)d\zeta = d\xi,$$

$$J_2 = \frac{\hat{a}y}{2\sqrt{\sigma\nu}}e^{-y\sqrt{\frac{\sigma}{\nu}}}\operatorname{erfc}\left(\frac{y}{2\sqrt{\nu t}} - \sqrt{\sigma t}\right) - \frac{\hat{a}y}{\sqrt{\pi\sigma\nu}}e^{-y\sqrt{\frac{\sigma}{\nu}}}\int_{\frac{y}{2\sqrt{\nu t}}}^{\infty} e^{-\left(\zeta - \frac{y}{2\zeta}\sqrt{\frac{\sigma}{\nu}}\right)^2}d\zeta. \tag{7.6}$$

Summing (7.5) and (7.6), we obtain

$$J_2 = \frac{\hat{a}y}{4\sqrt{\sigma\nu}}\left[e^{-y\sqrt{\frac{\sigma}{\nu}}}\operatorname{erfc}\left(\frac{y}{2\sqrt{\nu t}} - \sqrt{\sigma t}\right) - e^{y\sqrt{\frac{\sigma}{\nu}}}\operatorname{erfc}\left(\frac{y}{2\sqrt{\nu t}} + \sqrt{\sigma t}\right)\right].$$

The final expression for $\hat{V}$ is

$$\begin{aligned}\hat{V} &= \frac{1}{2}\hat{a}\left(t + \frac{y}{2\sqrt{\sigma\nu}}\right)e^{y\sqrt{\frac{\sigma}{\nu}}}\operatorname{erfc}\left(\frac{y}{2\sqrt{\nu t}} + \sqrt{\sigma t}\right) \\ &\quad + \frac{1}{2}\hat{a}\left(t - \frac{y}{2\sqrt{\sigma\nu}}\right)e^{-y\sqrt{\frac{\sigma}{\nu}}}\operatorname{erfc}\left(\frac{y}{2\sqrt{\nu t}} - \sqrt{\sigma t}\right).\end{aligned} \tag{7.7}$$

The introduce the characteristic fluid time

$$t_c = \frac{y}{2\sqrt{\sigma\nu}};$$

then

$$\hat{V} = \frac{1}{2}\hat{a}(t + t_c)e^{y\sqrt{\frac{\sigma}{\nu}}}\operatorname{erfc}\left(\frac{y}{2\sqrt{\nu t}} + \sqrt{\sigma t}\right) + \frac{1}{2}\hat{a}(t - t_c)e^{-y\sqrt{\frac{\sigma}{\nu}}}\operatorname{erfc}\left(\frac{y}{2\sqrt{\nu t}} - \sqrt{\sigma t}\right). \tag{7.8}$$

The velocity field obtained above can be used to analyze an unsteady boundary layer on the porous surface of a moving body.

Taking into account the asymptotics of the special functions involved, we can obtain the velocity field of the fluid as $t \to \infty$:

$$\hat{V}(y,t) = \hat{a}(t - t_c)e^{-y\sqrt{\frac{\sigma}{\nu}}}.$$

Note that the velocity field of the viscous fluid increases with time, but it starts to increase at some moment of time rather than from the very beginning, namely, after a quarter period of the main rotation of the system as a solid body, that is, at

$$t = t_c = \frac{1}{4}T = \frac{1}{4\Omega}.$$

In other words, the fluid begins to accelerate with some delay.

Concluding this chapter, we emphasize that the found solution of the unsteady Navier–Stokes 3D equations with an arbitrary viscosity in the semi-infinite space above a plane wall rotating with an angular velocity nonperpendicular to its place, as well as the solution of a similar system in the presence of a homogeneous injection modeled by a normal velocity constant on the plate, is of fundamental importance in determining the influence the fluid viscosity on the dynamics of the rigid body containing the fluid. The method for taking into account viscosity suggested here is characterized by a high level of formalism and makes it possible to allow for nonlinear effects of energy dissipation in the cavity.

Chapter 4

Control of a rotating rigid body containing a viscous fluid

This chapter considers the rotational motion of a rigid body with a cavity containing a viscous fluid. We derive general equations for the perturbed motion of a rigid body with a fluid which take into account the energy dissipation in the cavity related to the boundary layer effect. The shape of the cavity is arbitrary.

Section 1 studies small oscillations of a viscous fluid completely filling a cavity in a rotating body and determine properties of the eigenvalues of the problem.

In Section 2 we derive equations for a weakly perturbed motion of a body with a cavity containing a viscous fluid. Using the flate plate model for a cavity with smooth walls, we obtain an infinite system of integro-differential equations describing the perturbed motion of a body with a fluid. In the case of an ideal fluid, the system of equations transforms into the equations obtained in [277].

In Section 3 we calculate the coefficients of inertial couplings between the motions of the rigid body and the wave motions of the fluid; for a cylindrical cavity, detailed calculation is performed. In Section 4 we solve a system of integro-differential equations describing the motion of a rigid body with a cavity containing a viscous fluid which is weakly perturbed with respect to uniform rotation. This system is solved by the Laplace transform method. In Section 5 the moment of the viscous friction forces acting from the fluid on the shell of the rigid body is found. Section 6 considers the stability of a freely rotating rigid body with a cavity containing a viscous fluid. A characteristic equation for the oscillations of a fluid in a cavity in a rotating body is composed and solved by methods of perturbation theory.

Section 7 studies the oscillations of a viscous fluid in a cavity in a rigid body performing librational motion. A part of the cavity is filled with a viscous fluid and the other part, with a gas at constant pressure. In addition, the cavity carries constructive elements, such as radial and annular edges. Using the flat plate model, we derive equations for the perturbed motion of a body with a fluid. The energy dissipation during an oscillation period is assumed to be small in comparison with the energy of the system.

In Section 8, using the Laplace integral transform, we write the system of integro-differential equations with singular kernels in the space of images in the form of an algebraic equation. We also find a relationship between the images of the angular velocity and the perturbing moment. Using the inverse Laplace transform, we reduce the system of integro-differential equations to an integral relation of the same form as relation (3.7) in the first chapter for an ideal fluid.

In Section 9 this integral relation is reduced to the standard form, i.e., written in the form of a linear system of tenth-order differential equations.

1 SMALL OSCILLATIONS OF A VISCOUS FLUID COMPLETELY FILLING A VESSEL

Consider the motion of a viscous incompressible fluid in a coordinate system attached to a uniformly rotating body. Let $\boldsymbol{\omega}_0 = \omega_0 \boldsymbol{k}$, and $\boldsymbol{n}$ be the outer normal to the surface S (see Fig. 4.1).

The Navier–Stokes equations linearized with respect to the uniform rotation have the form

$$\begin{cases} \dfrac{\partial V}{\partial t} + 2\omega_0(\boldsymbol{k} \times V) = -\nabla p + \nu \Delta V, \\ \operatorname{div} V = 0 \quad \text{on } D, \quad V = 0 \quad \text{on } S. \end{cases} \tag{1.1}$$

We seek a solution of problem (1.1) in the form

$$V = \boldsymbol{v}(x_1, x_2, x_3)e^{\lambda t}, \quad p = P(x_1, x_2, x_3)e^{\lambda t}.$$

Thus,

$$\begin{cases} \lambda \boldsymbol{v} + 2\omega_0(\boldsymbol{k} \times \boldsymbol{v}) = -\nabla P + \nu \Delta \boldsymbol{v}, \\ \operatorname{div} \boldsymbol{v} = 0 \quad \text{on } D, \quad \boldsymbol{v} = 0 \quad \text{on } S. \end{cases} \tag{1.2}$$

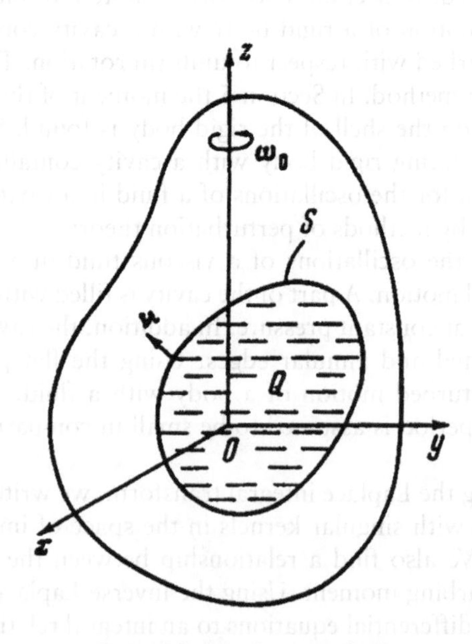

Figure 4.1 A body with a fluid.

Here the fluid occupies the volume D bounded by the surface of the walls of the cavity S and λ is a complex parameter.

Let $\boldsymbol{v}^+$ be the vector complex conjugate to $\boldsymbol{v}$. Multiplying the first equation in (1.2) by $\boldsymbol{v}^+$, we obtain

$$\begin{aligned} &\lambda(\boldsymbol{v},\boldsymbol{v}^+) + 2\omega_0(\boldsymbol{k}\times\boldsymbol{v})\boldsymbol{v}^+ = -(\nabla P,\boldsymbol{v}^+) + \nu(\boldsymbol{v}^+,\boldsymbol{\Delta v}), \\ &\operatorname{div}\boldsymbol{v} = 0 \quad \text{on } D, \quad \boldsymbol{v} = 0 \quad \text{on } S. \end{aligned} \tag{1.3}$$

Let us multiply the complex conjugate of (1.2) by $\boldsymbol{v}$:

$$\begin{aligned} &\lambda(\boldsymbol{v}^+,\boldsymbol{v}) + 2\omega_0(\boldsymbol{k}\times\boldsymbol{v}^+)\boldsymbol{v} = -(\nabla P^+,\boldsymbol{v}) + \nu(\boldsymbol{v},\boldsymbol{\Delta v}^+), \\ &\operatorname{div}\boldsymbol{v}^+ = 0 \quad \text{in } D, \quad \boldsymbol{v}^+ = 0 \quad \text{on } S. \end{aligned} \tag{1.4}$$

Summing (1.3) and (1.4), we obtain

$$(\lambda+\lambda^+)(\boldsymbol{v},\boldsymbol{v}^+) = -[(\boldsymbol{v}^+,\nabla P) + (\boldsymbol{v},\nabla P^+)] + \nu[(\boldsymbol{v},\boldsymbol{\Delta v}^+) + (\boldsymbol{v}^+,\boldsymbol{\Delta v})]. \tag{1.5}$$

Using the continuity equation, we arrive at

$$2\operatorname{Re}\lambda(\boldsymbol{v},\boldsymbol{v}^+) = -[\nabla(\boldsymbol{v}^+P + \boldsymbol{v}P^+)] - \nu[\boldsymbol{v}(\nabla\times\nabla\times\boldsymbol{v}^+) + \boldsymbol{v}^+(\nabla\times\nabla\times\boldsymbol{v})]. \tag{1.6}$$

Using the relations

$$\begin{aligned} &(\boldsymbol{v}^+,\boldsymbol{\Delta v}) = \operatorname{div}(\boldsymbol{v}^+\times\operatorname{rot}\boldsymbol{v}) - |\operatorname{rot}\boldsymbol{v}|^2, \\ &(\boldsymbol{v},\boldsymbol{\Delta v}^+) = \operatorname{div}(\boldsymbol{v}\times\operatorname{rot}\boldsymbol{v}^+) - |\operatorname{rot}\boldsymbol{v}|^2, \end{aligned}$$

we obtain

$$2\operatorname{Re}\lambda(\boldsymbol{v},\boldsymbol{v}^+) = -\operatorname{div}(\boldsymbol{v}^+P + \boldsymbol{v}P^+) + \nu[\operatorname{div}(\boldsymbol{v}^+\times\operatorname{rot}\boldsymbol{v} + \boldsymbol{v}\times\operatorname{rot}\boldsymbol{v}^+) - 2|\operatorname{rot}\boldsymbol{v}|^2]. \tag{1.7}$$

Let us integrate (1.7) over the volume D and apply Gauss' theorem:

$$\begin{aligned} 2\operatorname{Re}\lambda\int_D|\boldsymbol{v}|^2dD &= -\int_S[(\boldsymbol{v}^+,\boldsymbol{n})P + (\boldsymbol{v},\boldsymbol{n})P^+)]dS \\ &\quad + \nu\int_S[\boldsymbol{v}^+(\operatorname{rot}\boldsymbol{v}\times\boldsymbol{n}) + \boldsymbol{v}(\operatorname{rot}\boldsymbol{v}^+\times\boldsymbol{n})]dS - 2\nu\int_D|\operatorname{rot}\boldsymbol{v}|^2dD. \end{aligned} \tag{1.8}$$

The integrals of the first two terms over the surface vanish by virtue of the boundary conditions. Thus, we have

$$\operatorname{Re}\lambda = -\nu\frac{\int_D|\operatorname{rot}\boldsymbol{v}|^2dD}{\int_D|\boldsymbol{v}|^2dD}. \tag{1.9}$$

Similarly,

$$\operatorname{Im}\lambda = 2\omega_0 \frac{\int_D (\boldsymbol{v}\times\boldsymbol{v}^+)\boldsymbol{k}\,dD}{\int_D |\boldsymbol{v}|^2 dD}. \tag{1.10}$$

Let $\boldsymbol{v} = \boldsymbol{a} + i\boldsymbol{b}$, where $\boldsymbol{a}$ and $\boldsymbol{b}$ are real-valued vector functions; then

$$\begin{aligned} &\boldsymbol{v}\times\boldsymbol{v}^+ = (\boldsymbol{a}+i\boldsymbol{b})\times(\boldsymbol{a}-i\boldsymbol{b}) = -2i(\boldsymbol{a}\times\boldsymbol{b}), \\ &|\boldsymbol{k}(\boldsymbol{v}\times\boldsymbol{v}^+)| \leqslant 2|\boldsymbol{a}\times\boldsymbol{b}| \leqslant 2|\boldsymbol{a}|\cdot|\boldsymbol{b}| \leqslant |\boldsymbol{a}|^2+|\boldsymbol{b}|^2 = |\boldsymbol{v}|^2. \end{aligned} \tag{1.11}$$

Note that if the boundary S of the domain D is smooth, $\operatorname{div} v = 0$ on D, and the velocity component tangent to S vanishes on S, then

$$\int_D |\operatorname{rot}\boldsymbol{v}|^2 dD \geqslant C\int_D |\boldsymbol{v}|^2 dD.$$

These conditions hold in the case of a viscous fluid. The constant $C > 0$ depends only on the domain D [30].

Using (1.11) and the above inequality, we obtain

$$|\operatorname{Im}\lambda| \leqslant 2\omega_0, \quad \operatorname{Re}\lambda \leqslant -C\nu, \quad |\operatorname{Re}\lambda| \geqslant C\nu. \tag{1.12}$$

Thus, we have shown that $\operatorname{Im}\lambda$ densely fills the interval $[-2\omega_0, 2\omega_0]$, and the real part is bounded above by a negative constant.

Together with problem (1.2), we shall consider the corresponding problem for an ideal fluid, which is

$$\begin{cases} \lambda^0\boldsymbol{v}^0 + 2(\boldsymbol{\omega}_0\times\boldsymbol{v}^0) = -\nabla p^0, \\ \operatorname{div}\boldsymbol{v}^0 = 0 \quad \text{on } D, \quad (\boldsymbol{v}^0,\boldsymbol{n}) = 0 \quad \text{on } S. \end{cases} \tag{1.13}$$

In what follows, we assume the solution of problem (1.13) to be known. For a number of cavities, it was solved in [148, 299] and [58, 277].

If the viscosity of the fluid is low ($\nu \ll 1$), then the motion of the bulk of the fluid does not differ from the motion of an ideal fluid, except in a thin layer near the walls. In what follows, we are interested in the corrections to the frequency parameter λ determined by the influence of the boundary layer.

We shall solve the problem up to terms of order $O(\sqrt{\nu})$. Using the boundary layer technique, we represent the velocity and pressure field in the form

$$\begin{aligned} &\boldsymbol{v} = \boldsymbol{v}^0 + \sqrt{\nu}\boldsymbol{v}^1 + \cdots + \boldsymbol{W}, \\ &p = p^0 + \sqrt{\nu}p^1 + \cdots + q, \\ &\lambda = \lambda^0 + \sqrt{\nu}\lambda^1 + \cdots. \end{aligned} \tag{1.14}$$

Here $\boldsymbol{W}$ and q are functions of boundary layer type, which rapidly decrease with increasing the distance to the walls of the cavity. They can be expanded in asymptotic series in powers of $\sqrt{\nu}$. This yields the following boundary value problems for determining the unknowns functions:

$$(A)\quad \begin{cases} \lambda^0 \boldsymbol{v}^0 + 2(\boldsymbol{\omega}_0 \times \boldsymbol{v}^0) = -\nabla p^0, \\ \operatorname{div} \boldsymbol{v}^0 = 0 \quad \text{on } D, \quad (\boldsymbol{v}^0, \boldsymbol{n}) = 0 \quad \text{on } S. \end{cases}$$

$$(B)\quad \begin{cases} \lambda^0 \boldsymbol{v}^1 + 2(\boldsymbol{\omega}_0 \times \boldsymbol{v}^1) = -\nabla p^1 - \lambda^1 \boldsymbol{v}^0, \\ \operatorname{div} \boldsymbol{v}^1 = 0 \quad \text{on } D, \quad (\boldsymbol{v}^1, \boldsymbol{n}) = -\dfrac{(\boldsymbol{W}, \boldsymbol{n})}{\sqrt{\nu}} \quad \text{on } S. \end{cases}$$

$$(C)\quad \begin{cases} \lambda^0 \boldsymbol{W} + 2(\boldsymbol{\omega}_0 \times \boldsymbol{W}) = -\nabla q + \nu \boldsymbol{\Delta} \boldsymbol{W}, \\ \operatorname{div} \boldsymbol{W} = 0 \quad \text{on } D, \quad \boldsymbol{W}_T = -\boldsymbol{v}^0 \quad \text{on } S; \end{cases}$$

we have $\boldsymbol{W} \to 0$ and $q \to 0$ outside the boundary layer region D_S.

Let us introduce a curvilinear coordinate system so that $\xi = 0$ on the surface S and $\zeta > 0$ in D_S. Let W_ξ, W_η, and W_ζ be the components of the vector $\boldsymbol{W}$ in this curvilinear coordinate system, and let $\boldsymbol{e}_1$, $\boldsymbol{e}_2$, and $\boldsymbol{n}$ be the basis vectors. Making the change of variables $\zeta = \sqrt{\nu}\alpha$, $W_\zeta = \sqrt{\nu} W_\alpha$ and performing the asymptotic processing of problem (S) (passing to the limit as $\nu \to 0$), we obtain the problem

$$(D)\quad \begin{cases} \lambda^0 W_\xi - 2 W_\eta (\boldsymbol{\omega}_0, \boldsymbol{n}) = \dfrac{\partial^2 W_\xi}{\partial \alpha^2}, \\ \lambda^0 W_\eta + 2 W_\xi (\boldsymbol{\omega}_0, \boldsymbol{n}) = \dfrac{\partial^2 W_\eta}{\partial \alpha^2}, \\ \dfrac{\partial q^0}{\partial \alpha} = -2(\omega_1 W_\eta - \omega_2 W_\xi), \\ W_\xi = -v_\xi^0, \quad W_\eta = -v_\eta^0 \quad \text{at } \alpha = 0, \\ W_\xi, W_\eta, W_\alpha, q \to 0 \quad \text{as } \alpha \to \infty. \end{cases}$$

The solution of problem (D) satisfying the boundary conditions has the form

$$\boldsymbol{W}_T = -\frac{1}{2} \boldsymbol{v}^0 (e^{\mu_1 \alpha} + e^{\mu_2 \alpha}) - \frac{1}{2} i (\boldsymbol{v}^0 \times \boldsymbol{n})(e^{\mu_1 \alpha} - e^{\mu_2 \alpha}), \tag{1.15}$$

where $\mu_{1,2} = \sqrt{\lambda^0 \pm 2i(\boldsymbol{\omega}_0, \boldsymbol{n})}$.

For μ_1 and μ_2 we take those branches of the root for which $\operatorname{Re} \mu_k \leqslant 0$ at $k = 1, 2$. Passing to the old variables, we obtain

$$\boldsymbol{W}_T = -\frac{1}{2} \boldsymbol{v}^0 (e^{\mu_1 \frac{\zeta}{\sqrt{\nu}}} + e^{\mu_2 \frac{\zeta}{\sqrt{\nu}}}) - \frac{1}{2} i (\boldsymbol{v}^0 \times \boldsymbol{n})(e^{\mu_1 \frac{\zeta}{\sqrt{\nu}}} - e^{\mu_2 \frac{\zeta}{\sqrt{\nu}}}). \tag{1.16}$$

Solution (1.16) coincides with solution (5.4) obtained in Section 5 of Chapter 3 for a flat wall performing longitudinal oscillations in the half-space occupied by a viscous fluid.

Substituting (1.16) into (1.9) and (1.10) and performing simple but cumbersome calculations, we obtain

$$\lambda = \lambda^0 - \frac{\sqrt{\nu}(1+i)A}{2\sqrt{2}} + O(\nu),$$

$$A = \left(\int_D (v_0, v_0^+)dD\right)^{-1} \times \int_S \Big[(v_0, v_0^+)\left(\sqrt{\gamma_0 + 2(\omega_0, n)} + \sqrt{\gamma_0 - 2(\omega_0, n)}\right) \quad (1.17)$$

$$+ in(v_0^+ \times v_0)\left(\sqrt{\gamma_0 + 2(\omega_0, n)} - \sqrt{\gamma_0 - 2(\omega_0, n)}\right)\Big] dS;$$

here $\gamma_0 = i\lambda^0$.

By way of example, let us calculate the viscous additions for a cylindrical cavity. Consider a cylinder with height $2h$ and radius $R = 1$. We place a cylindrical coordinate system at the center of the cavity, so that the equations for the top and the bottom have the form $x = \pm h$ and the equation for the side is $R = 1$.

The solution of problem (A) for a cylindrical cavity is known [148]:

$$v_{\theta n} = \frac{\chi_n}{2\omega_0(\chi_n^2 - 1)}[\chi_n n \times \nabla\varphi_n - i\nabla\varphi_n + i\chi_n^2 n(n, \nabla\varphi_n)],$$

$$\varphi_n = g_n(x, r)e^{i\theta} = \sin(k_l x)\frac{J_1(\xi_{lp}R)}{J_1(\xi_{lp})}e^{i\theta}, \quad (1.18)$$

$$k_l = \frac{n(2l+1)}{2h}, \quad \xi_{lp} = k_l\sqrt{\chi_{lp}^2 - 1}$$

$$(l = 0, 1, 2, \ldots, \quad p = 1, 2, 3, \ldots).$$

No harmonics symmetric in the x coordinate are excited. The index n ranges over all possible combinations of the numbers l and p of longitudinal and transverse harmonics. The quantity ξ_{lp} is the pth root of the equation:

$$\xi J_0(\xi) - \left[1 \pm \sqrt{\left(\frac{\xi}{k_l}\right)^2 + 1}\right] J_1(\xi) = 0.$$

The eigenvalues χ_{lp} are in the vicinity of the value

$$[h^2p^2/(2l+1) + 1]^{1/2}$$

and densely fill both the positive and the negative part $|\chi| > 1$ of the real axis. Substituting (1.18) into (1.17), we obtain

$$A = \frac{1}{h}(\chi_{lp}^2(k_l^2 + 1) - \chi_{lp})^{-1} \times \Big\{\sqrt{\gamma_0 + 2\omega_0}[(k_l^2 + 1)(\chi_{lp} - 1)^2 - 2(\chi_{lp} - 1)]$$

$$+ \sqrt{\gamma_0 - 2\omega_0}[(k_l^2 + 1)(\chi_{lp} + 1)^2 - 2(\chi_{lp} + 1)] + 2h\sqrt{\lambda_0}\frac{k_l^2 + 1}{k_l^2}\xi_{lp}\Big\}.$$

It is seen from the formula

$$\lambda = \lambda^0 - \sqrt{\nu}\frac{(1+i)}{2\sqrt{2}}A + O(\nu)$$

how the viscosity of the fluid shifts the eigenfrequencies by a value proportional to $\sqrt{\nu}$.

2 EQUATIONS OF THE PERTURBED MOTION OF A BODY WITH A CAVITY CONTAINING A VISCOUS FLUID

The heuristic method presented here is close to that extensively used by L. D. Landau. This is a simple and visual method allowing for the effect of the second mechanism of energy dissipation, which is particularly important for applications.

The point is that, in the motion of a body with a cavity filled with a viscous fluid, it is viscosity which exerts a substantial influence on the stability of the steady rotation; moreover, the influence of viscosity is fairly sophisticated: in some cases, it ensures the stabilization of rotation, while in other cases, it leads to the loss of stability.

There are two mechanisms of energy dissipation under the oscillations of a fluid in a cavity. One of them is related to the vortex formation on the walls of the cavity and to the further energy dissipation in a thin near-wall layer (this occurs in the case of a cavity with smooth walls and large Reynolds numbers), and the other, to the separation of powerful discrete vortices, which afterwards dissipate over the entire volume of the fluid (this occurs in the case of a cavity having constructive elements with acute edges). The latter effect is substantially nonlinear; usually it is at least two orders higher than the boundary layer effect.

All this requires including additional dissipative forces in the equations of motion of a rigid body with a fluid-containing cavity on which the study of stability is based.

This section considers a class of perturbed motions of a body with a fluid with small relative energy dissipation, as well as generalized coordinates characterizing the perturbed motion of the rigid body and the fluid. The shape of the cavity is arbitrary. We also reduce the infinite system of integro-differential equations obtained for the case of a cavity with smooth walls to a finite system of differential equations. These results contain results of [261, 277] as a special case.

1. Consider a rigid body with a cavity filled with a viscous incompressible fluid of density ρ and kinematic viscosity ν in the field of mass forces with potential U. The walls of the cavity are assumed to be smooth.

Let us write the equations of motion of the fluid in any coordinate system $Ox_1x_2x_3$ attached to the body:

$$\begin{cases} \boldsymbol{w}_0 + \boldsymbol{\omega} \times (\boldsymbol{\omega} \times \boldsymbol{r}) + \dot{\boldsymbol{\omega}} \times \boldsymbol{r} + 2\,\boldsymbol{\omega} \times \boldsymbol{V} + \dfrac{\partial \boldsymbol{V}}{\partial t} + (\boldsymbol{V}, \nabla)\boldsymbol{V} = -\dfrac{\nabla P}{\rho} - \nabla U + \nu \Delta \boldsymbol{V}, \\ \operatorname{div} \boldsymbol{V} = 0 \quad \text{on } Q, \\ \boldsymbol{V} = 0 \quad \text{on the surface } S, \quad \boldsymbol{V} = \boldsymbol{V}_0(\boldsymbol{r}) \quad \text{at } t = 0. \end{cases} \tag{2.1}$$

Here t is time, $\boldsymbol{r}$ is the radius vector from O, $\boldsymbol{V}$ is the velocity of the fluid in the coordinate system $Ox_1x_2x_3$, P is the pressure, $\boldsymbol{w}_0$ is the absolute acceleration of the point O, ω si the absolute angular velocity of the body, $\dot{\omega}$ is its angular acceleration, S is the boundary of the domain Q, and n is the unit outer normal vector to S.

In the coordinate system $Ox_1x_2x_3$, the equations of moments about the center of inertia O_1 of the entire system has the form

$$\frac{dK}{dt} + \omega \times K = M_1, \quad K = J\omega + \rho \int_Q r \times V dQ, \tag{2.2}$$

where K is the kinetic momentum.

We assume that the unperturbed motion is the uniform rotation of the body with a fluid about the axis O_1y passing through the center of inertia O_1 of the whole system and parallel to the Ox_3 axis at an angular velocity $\omega_0 = \omega_0 k$, where k is the basis vector along the Ox_3 axis.

For the unperturbed motion, we have

$$V = 0, \quad M_1 = \omega_0 \times J\omega_0,$$

where M_1 is the resultant moment of external forces about the center of inertia O_1 and J is the inertia tensor of the entire system with respect to the point O_1, which is composed of the inertia tensor I^0 of the rigid body and the inertia tensor I of the fluid with respect to the same point. A body with a fluid-filled cavity is a gyrostat. Therefore, the center of inertia O_1 of the system is immovable with respect to the body coordinate system $Ox_1x_2x_3$, and the tensors I^0, I, and J are constant in this coordinate system.

We characterize the perturbed motion of a rigid body with a cavity containing a viscous fluid by the vector $\Omega(t)$ so that the angular velocity of the body is represented in the form $\omega = \omega_0 + \Omega(t)$ and the angular velocity $\Omega(t)$ is first-order small in comparison with ω_0.

We set

$$P = \rho\left[p + \frac{1}{2}(\omega \times r)^2 - U - (w_0, r)\right], \tag{2.3}$$

$$M_1 = \omega_0 \times J\omega_0 + M$$

and assume the quantities V, p, and M to be first-order small. Here M is the moment of the external forces acting on the rigid body (without fluid).

Neglecting the terms of higher order of smallness with respect to the generalized coordinates, we can write the equations of motion for the fluid in the form

$$\begin{cases} \dfrac{\partial V}{\partial t} + 2\,\omega_0 \times V + \dot{\Omega} \times r = -\nabla p + \nu \Delta V, \\ \operatorname{div} V = 0 \quad \text{on } Q, \quad V = 0 \quad \text{on } S, \quad V = V_0(r) \quad \text{at } t = 0. \end{cases} \tag{2.4}$$

Similarly, the equations of motion of the body with a fluid can be written as

$$(I+I^0)\dot{\boldsymbol{\Omega}}+\boldsymbol{\Omega}\times(I+I^0)\boldsymbol{\omega}_0+\boldsymbol{\omega}_0\times(I+I^0)\boldsymbol{\Omega}+\rho\int\limits_Q \boldsymbol{r}\times\dot{\boldsymbol{V}}dQ+\rho\int\limits_Q \boldsymbol{\omega}_0\times(\boldsymbol{r}\times\boldsymbol{V})dQ$$
$$=\boldsymbol{M}+\delta\boldsymbol{M}. \tag{2.5}$$

Here $\boldsymbol{M}$ is the moment of external forces and $\delta\boldsymbol{M}$ is the moment of generalized forces, which emerges because of the energy dissipation in the cavity.

Equations (2.4) and (2.5), together with the usual equations of motion of the center of inertia, the kinematic relations, and the initial conditions, completely describe the dynamics of a body with a fluid.

The hydrodynamic part of the problem reduces to solving the boundary eigenvalue problem

$$\begin{cases} \Delta\varphi+\sigma^2\dfrac{\partial^2\varphi}{\partial x_3^2}=0 \quad \text{on } Q, \\ \\ (L\nabla\varphi,\boldsymbol{n})=0 \quad \text{on } S, \end{cases} \tag{2.6}$$

where $L\boldsymbol{b}=\boldsymbol{b}+\sigma^2\boldsymbol{k}(\boldsymbol{k},\boldsymbol{b})+\sigma\,\boldsymbol{b}\times\boldsymbol{k}$ is the linear transformation introduced in [36] and $\boldsymbol{n}$ is the outer normal vector to S. As is known [299], problem (2.6) has countably many eigenfunctions φ_n and eigenvalues σ_n, which densely fill the domain $\operatorname{Re}\sigma_n=0$, $|\sigma_n|\geqslant 1$. The complex vector functions $\boldsymbol{V}_n(x_1,x_2,x_3)$ determined by

$$\boldsymbol{V}_n=-\lambda_n^{-1}(1+\sigma_n^2)^{-1}L\nabla\varphi_n, \quad \lambda_n=2\omega_0/\sigma_n, \tag{2.7}$$

are orthogonal in Q [277] and satisfy the equations

$$\begin{cases} \lambda_n\boldsymbol{V}_n+2\,\boldsymbol{\omega}_0\times\boldsymbol{V}_n+\nabla\varphi_n=0, \\ \operatorname{div}\boldsymbol{V}_n=0 \quad \text{on } Q, \quad (\boldsymbol{V}_n,\boldsymbol{n})=0 \quad \text{on } S. \end{cases} \tag{2.8}$$

We seek solutions of equations (2.4) by the Galerkin method in the form of series with coefficients S_n and u_n:

$$\boldsymbol{V}=\sum_{n=1}^{\infty}S_n(t)\boldsymbol{V}_n(x_1,x_2,x_3),$$
$$p=\sum_{n=1}^{\infty}u_n(t)\varphi_n(x_1,x_2,x_3). \tag{2.9}$$

We represent the term $\dot{\boldsymbol{\Omega}}\times\boldsymbol{r}$ in (2.4), which describes the motion of the body, in the form

$$\dot{\boldsymbol{\Omega}}\times\boldsymbol{r}=\sum_{n=1}^{\infty}\mu_n^{-1}(\boldsymbol{a}_n^*,\dot{\boldsymbol{\Omega}})\boldsymbol{V}_n, \quad \boldsymbol{a}_n=\int\limits_Q \boldsymbol{r}\times\boldsymbol{V}_n\,dQ, \tag{2.10}$$

where $\mu_n=\rho\int\limits_Q|\boldsymbol{V}_n|^2dQ$.

Let us substitute (2.9) and (2.10) into (2.4), replacing $2\,\omega_0 \times V_n$ by its expression

$$2\,\omega_0 \times V_n = -\lambda_n V_n - \nabla\varphi_n.$$

Applying the Galerkin procedure, we obtain the following equations for the generalized Fourier coefficients of the velocity:

$$\begin{aligned} &\mu_n[\dot{S}_n(t) - \lambda_n S_n(t)] + (a_n^*, \dot{\boldsymbol{\Omega}}) = \delta P_n, \\ &S_n = S_{n0} \quad \text{at } t = 0 \ (n = 1, 2, \ldots). \end{aligned} \tag{2.11}$$

Here the δP_n are the generalized forces caused by the energy dissipation in the cavity. The asterisk denotes complex conjugation.

Taking into account (2.9) and (2.10), we write the kinetic momentum $\boldsymbol{K}$ in the form

$$\boldsymbol{K} = J\omega_0 + J\boldsymbol{\Omega} + \sum_{n=1}^{\infty} S_n(t) a_n. \tag{2.12}$$

Substituting (2.12) into (2.5), we obtain the following equations of motion for the body with a fluid:

$$J\dot{\boldsymbol{\Omega}} + \boldsymbol{\Omega} \times J\omega_0 + \omega_0 \times J\boldsymbol{\Omega} + \sum_{n=1}^{\infty} [a_n \dot{S}_n + (\omega_0 \times a_n) S_n] = \boldsymbol{M} + \delta\boldsymbol{M}. \tag{2.13}$$

Equations (2.11) and (2.13), together with the initial conditions for $\boldsymbol{\Omega}(t)$, completely determine the dynamics of a body with a fluid.

Thus, the problem of the dynamics of a rotating body with a fluid-containing cavity decomposes into two parts, which can be considered independently. The first, hydrodynamic, part of the problem reduces to solving the boundary value problem (2.6); it depends only on the geometry of the cavity and does not depend on the motion of the body. It requires calculating the coefficients μ_n and a_n.

The second, dynamic, part of the problem reduces to solving the Cauchy problem for the system of ordinary differential equations (2.13), which can be done by well-known methods analytically or numerically.

For $\delta\boldsymbol{M} = \delta P_n = 0$, equations (2.2) and (2.4) transform into the equations of the perturbed motion of a body with a cavity filled with an ideal fluid, which is the vector counterpart of the equations derived in [221], where explicit expressions for the corresponding coefficients are given. In what follows, we assume that the coefficients in the equations of motion of a body with an ideal fluid are known.

Thus, the problem of deriving equations of motion for a body with a cavity completely filled with a viscous fluid reduces to determining the generalized dissipative forces $\delta\boldsymbol{M}$ and δP_n.

2. Consider the case of a cavity with smooth walls and the motion of the fluid at large Reynolds numbers $\mathrm{Re} \gg 1$.

In this case, energy dissipates in a thin boundary layer near the boundary of the region occupied by the fluid. The rate of energy dissipation per unit volume of the fluid is of the order of $1/\mathrm{Re}$ near the wetted surface S of the cavity. In what follows, we leave only terms of the order of $1/\mathrm{Re}$ in expressions for generalized dissipative forces. Since the thickness of the boundary layer is small (of the order of $1/\mathrm{Re}$), an element of the surface S can be identified with an element of the infinite plane bounding the half-space filled with a viscous fluid moving at a velocity equal to the difference of the velocities of the rigid body and the perturbed motion of the ideal fluid. The force acting on an element of the surface S moving at a velocity $\boldsymbol{V}(\boldsymbol{r},t)$ can be written in the form (see (4.4) in Chapter 3)

$$dF = -\rho\sqrt{\frac{\nu}{\pi}} \times \left\{ \int_0^t \frac{\frac{\partial}{\partial\tau}[\boldsymbol{V}(\boldsymbol{r},t)\cos 2(\boldsymbol{\omega}_0,\boldsymbol{n})(t-\tau) + \boldsymbol{V}(\boldsymbol{r},t)\times\boldsymbol{n}\sin 2(\boldsymbol{\omega}_0,\boldsymbol{n})(t-\tau)]}{\sqrt{t-\tau}}d\tau \right\} dS, \tag{2.14}$$

where $\boldsymbol{r}$ is the radius vector of the current point of S in the coordinate system $Ox_1x_2x_3$. Thus, the rate of energy dissipation in the cavity equals

$$\frac{dE}{dt} = \int_S (\boldsymbol{V}, dF). \tag{2.15}$$

We represent the velocity of the perturbed motion of the ideal fluid in the form [221]

$$\boldsymbol{V}^0 = \sum_{k=1}^{\infty} S_k(t)\boldsymbol{V}_k(\boldsymbol{r}) + \sum_{j=1}^{3} \Omega_j \nabla\Psi_j, \tag{2.16}$$

where the $\Omega_j = (\boldsymbol{\Omega}, \boldsymbol{e}_j)$ are the components of the vector $\boldsymbol{\Omega}$ in the coordinate system $Ox_1x_2x_3$, the $\boldsymbol{e}_j$ are the basis vectors of the system $Ox_1x_2x_3$, and the $\Psi_j(x_1,x_2,x_3)$ and $\varphi_j(x_1,x_2,x_3)$ are functions being solutions of boundary value problems depending on the geometry of the cavity.

For the relative velocity $\boldsymbol{V}(\boldsymbol{r},t)$, we obtain the expression

$$\boldsymbol{V} = \Omega \times \boldsymbol{r} - \boldsymbol{V}^0 = -\sum_{j=1}^{3} \Omega_j(\nabla\Psi_j + \boldsymbol{r}\times\boldsymbol{e}_j) - \sum_{k=1}^{\infty} S_k(t)\boldsymbol{V}_k(\boldsymbol{r}).$$

Substituting this expression into (2.15), we obtain

$$\dot{E} = -\sqrt{\frac{\nu}{\pi}} \sum_{m=1}^{\infty} S_m(t) \sum_{k=1}^{\infty} \int_0^t \frac{\dot{S}_k(\tau)\alpha_{mk}(t-\tau) + S_k(\tau)\beta_{mk}(t-\tau)}{\sqrt{t-\tau}} d\tau, \tag{2.17}$$

where

$$\alpha_{mk} = \rho \int_S (V_m^*, [V_k \cos(2\omega_0, n)(t-\tau) + V_k \times n \sin(2\omega_0, n)(t-\tau)])dS$$

$$+ \rho \int_S (r \times e_j + \nabla\Psi_j, V_m^*)dS, \tag{2.18}$$

$$\beta_{mk} = \rho \int_S (2\omega_0, n)(V_m^*, [V_k \sin(2\omega_0, n)(t-\tau) - V_k \times n \cos(2\omega_0, n)(t-\tau)])dS$$

$$+ \rho \int_S (2\omega_0, n)(r \times e_j + \nabla\Psi_j, V_m^*)dS \quad (j = 1, 2, 3, \ m, k = 1, 2, \ldots). \tag{2.19}$$

On the other hand, the rate of energy dissipation is determined by the expression

$$\dot{E} = (\delta M, \Omega) + \sum_{m=1}^{\infty} \delta P_m S_m. \tag{2.20}$$

Comparing (2.17) and (2.20), we find

$$\delta M = 0,$$

$$\delta P_m = -\sqrt{\frac{\nu}{\pi}} \sum_{k=1}^{\infty} \int_0^t \frac{\dot{S}_k(\tau)\alpha_{mk}(t-\tau) + S_k(\tau)\beta_{mk}(t-\tau)}{\sqrt{t-\tau}} d\tau. \tag{2.21}$$

Substituting the expressions (2.21) for generalized forces into equations (2.2) and (2.4), we obtain the following equation for the perturbed motion of a body with a cavity filled with a viscous fluid at large Reynolds numbers:

$$J\dot{\Omega} + \Omega \times J\omega_0 + \omega_0 \times J\Omega + \sum_{n=1}^{\infty} [a_n \dot{S}_n + (\omega_0 \times a_n)S_n] = M,$$

$$\mu_n \left\{ \dot{S}_n - i\lambda_n S_n + \sqrt{\frac{\nu}{\pi}} \int_0^t \frac{\dot{S}_n(\tau)\alpha_n(t-\tau) + S_n(\tau)\beta_n(t-\tau)}{\sqrt{t-\tau}} d\tau \right\} + (a_n^*, \dot{\Omega}) \tag{2.22}$$

$$= -\sqrt{\frac{\nu}{\pi}} \sum_{m=1}^{\infty} \int_0^t \frac{\dot{S}_n(\tau)\alpha_{mn}(t-\tau) + S_n(\tau)\beta_{mn}(t-\tau)}{\sqrt{t-\tau}} d\tau,$$

where $\alpha_n = \alpha_{nn}/\mu_n$ and $\beta_n = \beta_{nn}/\mu_n$ for $n = 1, 2, \ldots$.

The coefficients β_n, β_{mn}, α_n, and α_{mn} in the infinite system of integro-differential equations (2.22), which are related to energy dissipation in the cavity, are determined by relations (2.18) and (2.19) and can be expressed in terms of solutions of the same boundary value problems as the coefficients in the equations of motion of a body with an ideal fluid given in [221].

The system of equations (2.22) can be regarded as Lagrange equations of the second kind in which the Ω_j and the S_k (where $j = 1,2,3$ and $k = 1,2,\ldots$) play the role of generalized coordinates. The system of equations (2.22) makes it possible to consider regimes in which the body is subject to the action of any control moments.

Problems of the dynamics of rotational motions of a body with a fluid-containing cavity were considered by various authors. The study of the dynamics of a rotating fluid puts forth a number of complicated problems of purely mathematical character. Some of them have been considered, but a number of questions, especially those related to the case of a viscous fluid, require further study. The purpose of this chapter is to decompose, whenever possible, the problem of the motion of a body with a fluid into hydrodynamic and dynamic parts. The former reduces to calculating certain functions depending on the shape of the cavity and tensors expressed in terms of these function. The latter, which consists in studying the motion of the body, uses only those tensors which characterize the action of the fluid on the body.

3 COEFFICIENTS OF INERTIAL COUPLINGS OF A RIGID BODY WITH A FLUID: THE CASE OF A CYLINDRICAL CAVITY

In this section we calculate coefficients characterizing the interaction between the motion of the rigid body and the wave motions of the fluid, namely, the coefficients α_{mn}, β_{mn}, μ_n, and a_n.

According to equations (2.6) of the preceding section, the boundary value problem for the functions $\varphi_n(x_1,x_2,x_3)$ has the form

$$\begin{gathered}\frac{\partial^2\varphi_n}{\partial x_1^2}+\frac{\partial^2\varphi_n}{\partial x_2^2}+(1-\chi_n^2)\frac{\partial^2\varphi_n}{\partial x_3^2}=0 \quad \text{on } Q,\\ \frac{\partial\varphi_n}{\partial x_1}n_1+\frac{\partial\varphi_n}{\partial x_2}n_2+(1-\chi_n^2)\frac{\partial\varphi_n}{\partial x_3}n_3=i\chi_n\left(\frac{\partial\varphi_n}{\partial x_2}n_1-\frac{\partial\varphi_n}{\partial x_1}n_2\right) \quad \text{on } S.\end{gathered} \tag{3.1}$$

Here n_1, n_2, and n_3 are the direction cosines of the unit outer normal $\boldsymbol{n}$ to the surface S and $\chi_n = i\sigma_n$.

The vector functions $V_n(x_1,x_2,x_3)$ defined by

$$V_n=\frac{\chi_n}{2\omega_0(\chi_n^2-1)}[\chi_n\boldsymbol{k}\times\nabla\varphi_n-i\nabla\varphi_n+i\chi_n^2\boldsymbol{k}(\boldsymbol{k},\nabla\varphi_n)], \tag{3.2}$$

are orthogonal in Q. According to relations (2.6), (2.18), and (2.19) of the preceding section, we write expressions for the hydrodynamic coefficients as

$$\mu_n = \rho \int_Q (\boldsymbol{V}_n, \boldsymbol{V}_n^*) dQ, \quad \boldsymbol{a}_n = \int_Q \boldsymbol{r} \times \boldsymbol{V}_n \, dQ,$$

$$\alpha_{mn} = \rho \int_S (\boldsymbol{V}_m^*, [\boldsymbol{V}_n \cos(2\boldsymbol{\omega}_0, \boldsymbol{n})(t-\tau) + \boldsymbol{V}_n \times \boldsymbol{n} \sin(2\boldsymbol{\omega}_0, \boldsymbol{n})(t-\tau)]) dS, \tag{3.3}$$

$$\beta_{mn} = \rho \int_S (2\boldsymbol{\omega}_0, \boldsymbol{n})(\boldsymbol{V}_m^*, [\boldsymbol{V}_n \sin(2\boldsymbol{\omega}_0, \boldsymbol{n})(t-\tau) - \boldsymbol{V}_n \times \boldsymbol{n} \cos(2\boldsymbol{\omega}_0, \boldsymbol{n})(t-\tau)]) dS$$

$$(m, n = 1, 2, \dots).$$

In the case of an axially symmetric cavity, the hydrostatic moment of the ideal fluid with respect to the axis of symmetry vanishes, and the scalar equation of motion about the Ox_3 axis can be separated from the other equations. The equations of rotation of the body about the transverse axes Ox_1 and Ox_2 are identical.

We introduce the cylindrical coordinate system

$$x_1 = R\cos\theta, \quad x_2 = R\sin\theta, \quad x_3 = z.$$

Let us represent the functions $\varphi_n(x_1, x_2, x_3)$ as the products

$$\varphi_n = g_n(z, R)e^{i\theta}.$$

This representation is natural, because the coefficients $\boldsymbol{a}_n$ of the other harmonics in the circular coordinate θ vanish, i.e., are not excited in the mechanical system under consideration [57]. According to (3.1), each function $g_n(z, R)$ must satisfy the two-dimensional boundary value problem with real coefficients

$$\frac{\partial^2 g_n}{\partial R^2} + \frac{1}{R}\frac{\partial g_n}{\partial R} - \frac{g_n}{R^2} + (1 - \chi_n^2)\frac{\partial^2 g_n}{\partial z^2} = 0,$$
$$\left(\frac{\partial g_n}{\partial R} + \chi_n \frac{g_n}{R}\right) n_R + (1 - \chi_n^2)\frac{\partial g_n}{\partial z} n_z = 0. \tag{3.4}$$

The first equation in (3.4) must be satisfied in the meridional section G of the cavity, and the second, on the boundary of this section. Here $\boldsymbol{n} = (n_R, n_\theta, n_z)$ is the unit outer normal vector to the surface S.

Suppose that the cavity Q is a cylinder of unit radius with height $2h$, that is,

$$Q = \{(R, \theta, z), \ 0 \leqslant R \leqslant 1, \ 0 \leqslant \theta \leqslant 2\pi, \ -h \leqslant z \leqslant h\}.$$

The solution of the boundary value problem (3.4) is known [57]:

$$g_n = g_{lp}(z,R) = \sin(k_l z)\frac{J_1(\xi_{lp}R)}{J_1(\xi_{lp})},$$
$$k_l = \frac{\pi(2l+1)}{2h}, \quad \xi_{lp} = k_l\sqrt{\chi_{lp}^2 - 1}, \quad (l = 0,1,\ldots, \ p = 1,2,\ldots). \tag{3.5}$$

Harmonics symmetric with respect to the coordinate z are not excited. The index n ranges over all combinations of the numbers l and p of the longitudinal and transverse harmonic. The quantity ξ_{lp} is the pth root of the equation

$$\xi J_1'(\xi) + \chi J_1(\xi) = 0 \quad \text{or} \quad \xi J_0(\xi) - \left[1 \pm \sqrt{\left(\frac{\xi}{k_l}\right)^2 + 1}\right] J_1(\xi) = 0. \tag{3.6}$$

where $J_0(\xi)$ and $J_1(\xi)$ are the Bessel function.

We represent the formula (3.3) for calculating the coefficients α_{mn} as the sum of two integrals:

$$\alpha_{mn} = J_1^{(m,n)} + J_2^{(m,n)},$$

where

$$J_1^{(m,n)} = \rho \int_S (\boldsymbol{V}_m^*, \boldsymbol{V}_n)\cos(2\boldsymbol{\omega}_0, \boldsymbol{n})(t-\tau)dS,$$
$$J_2^{(m,n)} = \rho \int_S (\boldsymbol{V}_m^*, \boldsymbol{V}_n \times \boldsymbol{n})\sin(2\boldsymbol{\omega}_0, \boldsymbol{n})(t-\tau)dS. \tag{3.7}$$

In these integrals, $m = (q,s)$, $n = (l,p)$, and the velocity field is determined by (3.2).

Calculating the gradient of the functions φ_n in the cylindrical coordinate system, determining the velocity field of the fluid by (3.2), and substituting the results into (3.7), we find $J_1^{(m,n)}$ and $J_2^{(m,n)}$.

Since the eigenfunctions of the boundary value problem (3.1) are orthogonal, it follows that the integrals $J_1^{(m,n)}$ and $J_2^{(m,n)}$ vanish at $m \neq n$. Thus, these integrals are calculated for $s = l$ and $q = p$; i.e., they have the forms $J_1^{(n,n)}$ and $J_2^{(n,n)}$, where $n = (l,p)$. We denote these integrals by $J_1^{(l,p)}$ and $J_2^{(l,p)}$, respectively. A simple but cumbersome calculation reduce the expression for α_{lp} to the form

$$\alpha_{lp} = \frac{\pi\rho\chi_n^2}{2\omega_0^2(\chi_n^2 - 1)}[A_n + B_n\cos 2\omega_0(t-\tau) + 2iC_n\sin 2\omega_0(t-\tau)], \tag{3.8}$$

where

$$A_n = h(1 + k_l^2)(\chi_n^2 - 1),$$
$$B_n = k_l^2(\chi_n^2 + 1) + (\chi_n - 1)^2,$$
$$C_n = k_l^2 \chi_n + (\chi_n - 1), \quad n = (l, p).$$

We represent the formula (3.3) for calculating the coefficients β_{mn} in the form

$$\beta_{mn} = I_1^{(m,n)} + I_2^{(m,n)},$$

where

$$I_1^{(m,n)} = \rho \int_S (2\omega_0, \boldsymbol{n}) \left(\mathbf{V}_m^*, \mathbf{V}_n\right) \sin(2\omega_0, \boldsymbol{n})(t - \tau) dS,$$
$$I_2^{(m,n)} = \rho \int_S (2\omega_0, \boldsymbol{n}) \left(\mathbf{V}_m^*, \mathbf{V}_n \times \boldsymbol{n}\right) \cos(2\omega_0, \boldsymbol{n})(t - \tau) dS. \tag{3.9}$$

As mentioned at the beginning of this section, the eigenfunctions of the boundary value problem (3.1) involved in the integrals $I_1^{(m,n)}$ and $I_2^{(m,n)}$ are orthogonal on the surface S. Therefore, these integrals vanish if $m \neq n$, where $m = (q, s)$ and $n = (l, p)$.

Below we evaluate integrals of the form (3.9) for $s = l$ and $q = p$, i.e., determine the integrals $I_1^{(n,n)}$ and $I_2^{(n,n)}$, which we denote by $I_1^{(l,p)}$ and $I_2^{(l,p)}$, respectively. Performing a fairly cumbersome calculation, we obtain the following expression for the coefficients β_{lp}:

$$\beta_{lp} = \frac{\pi \rho \chi_n^2}{\omega_0^2(\chi_n^2 - 1)} [B_n \sin 2\omega_0(t - \tau) + 2iC_n \cos 2\omega_0(t - \tau)]. \tag{3.10}$$

Here the B_n and the C_n are the same coefficients as in formula (3.8) for $n = (l, p)$.

To determine the coefficients μ_{lp}, we return to formulas (3.3):

$$\mu_n = \rho \int_Q (\mathbf{V}_n, \mathbf{V}_n^*) dQ,$$

where $dQ = R\, dR\, d\theta\, dz$.

Integrating, we obtain

$$\mu_{lp} = \frac{\pi \rho h \chi_n^3}{2\omega_0^2(\chi_n^2 - 1)} C_n, \quad n = (l, p), \tag{3.11}$$

where the C_n are calculated by (3.8).

To determine the gyrostatic coefficients a_n, we write out the formula

$$a_n = \int_Q r \times V_n \, dQ$$

from (3.3), where dQ the volume element in the cylindrical coordinate system.

Calculating the vector product and integrating the resulting expression over the volume, we obtain

$$a_n = \frac{4\chi_n(-1)^l}{\omega_0 \xi_n^4} \times [-i(\chi_n \xi_n^2 - (\chi_n - 1)^2) e_R + ((2 + \chi_n)\xi_n^2 + (\chi_n^2 - 1)^2) e_\theta]. \quad (3.12)$$

In what follows, we use the complex conjugate coefficients

$$a_n^* = \frac{4\chi_n(-1)^l}{\omega_0 \xi_n^4} [i(\chi_n \xi_n^2 - (\chi_n - 1)^2) e_R + ((2 + \chi_n)\xi_n^2 + (\chi_n^2 - 1)^2) e_\theta].$$

The results of computation of the coefficients in the equations of perturbed motion for a cylindrical cavity are given in Figs. 4.2–4.5. Figures 4.2 and 4.3 present the eigenvalues χ_{lp} and the inertial coupling coefficients $E_{lp} = 2\rho a_{lp}^2 / \mu_{lp}^2$ as functions of l for various h at $\rho = 1$.

Figures 4.4 and 4.5 show the quantities χ_{lp} and E_{lp} as functions of height for various values of p at $l = 0$. The character of the behavior of the curves for the succeeding values $p \geqslant 2$ and $l \geqslant 1$ is similar to that for the preceding values.

The found inertial coupling coefficients α_{lp}, β_{lp}, μ_{lp}, and a_{lp} close the system of integro-differential equations describing the dynamics of a rigid body with a viscous fluid performing rotational motions.

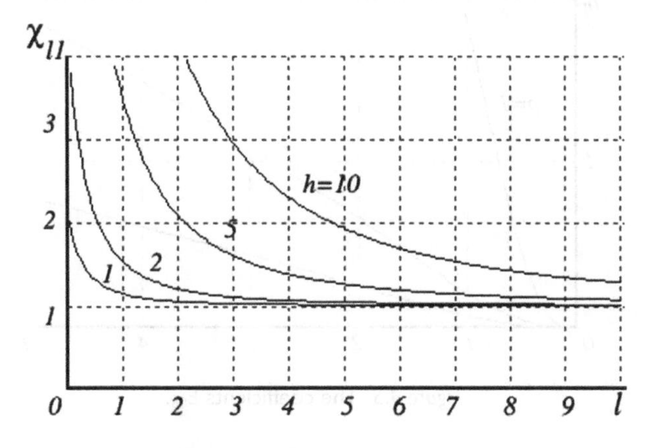

Figure 4.2 The frequencies χ_{l1}.

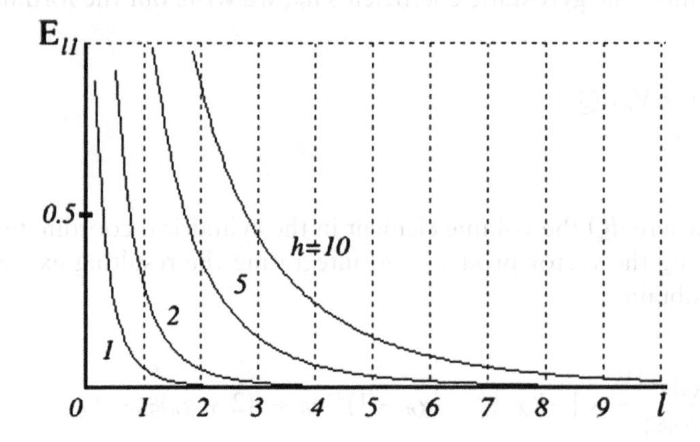

Figure 4.3 The coefficients E_{l1}.

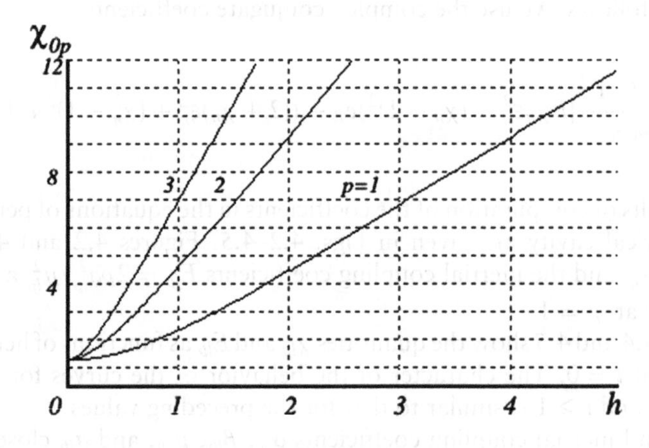

Figure 4.4 The frequencies χ_{0p}.

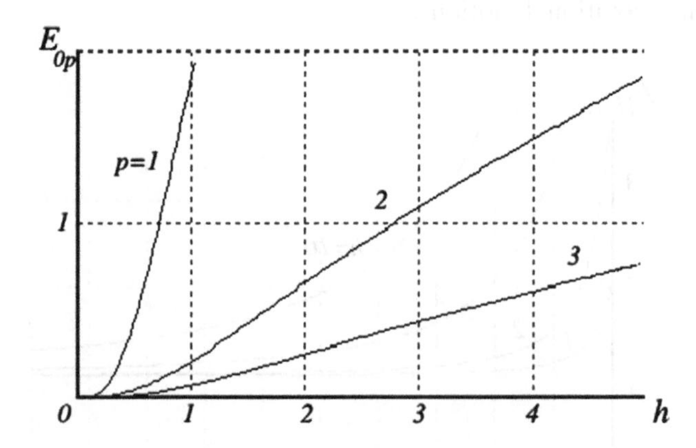

Figure 4.5 The coefficients E_{0p}.

4 OSCILLATIONS OF A VISCOUS INCOMPRESSIBLE FLUID IN A CAVITY OF A ROTATING BODY

This section considers a system of integro-differential equations describing the motion of a rigid body with a cavity containing a viscous fluid perturbed relative to uniform rotation. This system of equations is solved by using the Laplace transform. The obtained solutions can be used for analyzing and calculating the dynamics of spacecrafts with a fluid-filled cavity aboard.

In studying the oscillations of a rigid body with a cavity filled with a viscous fluid, we obtain general equations of motion for a body with a fluid. These equations are integro-differential.

To derive the equations of motion, we must determine the velocity field of the ideal fluid filling the cavity of the rotating body, i.e., determine the Zhukovskii potentials for the given cavity. This problem, which depends on the geometry of the cavity, is on the free oscillations of a fluid in an immovable vessel. Knowing the Zhukovskii potentials and the frequency of free oscillations of the fluid, we can calculate coefficients characterizing the influence of the fluid on the motion of the body.

The system of integro-differential equations describing the motion of a rigid body with a cavity containing a viscous fluid perturbed relative to uniform rotation is integrated by the Laplace method. Solving this system of equations, we obtain the amplitudes $s_n(t)$ of the nth tone of fluid oscillations and, thereby, determine the velocity field of the viscous incompressible fluid in the cavity of the rotating body. An important advantage of the solved system of equations is that its parameters are determined by characteristics of linear problems which have been calculated for a large class of cavities [36, 232].

In Section 2 of this chapter, we obtained the following equations of motion for a rigid body with a cavity containing a viscous fluid at large Reynolds numbers (2.22):

$$J\dot{\mathbf{\Omega}} + \mathbf{\Omega} \times J\boldsymbol{\omega}_0 + \boldsymbol{\omega}_0 \times J\mathbf{\Omega} + \sum_{n=1}^{\infty} [\boldsymbol{a}_n \dot{s}_n(t) + (\boldsymbol{\omega}_0 \times \boldsymbol{a}_n) s_n(t)] = \boldsymbol{M}(t), \tag{4.1}$$

$$\mu_n \dot{s}_n - i\lambda_n \mu_n s_n + (\boldsymbol{a}_n^*, \dot{\mathbf{\Omega}}) + \sqrt{\frac{\nu}{\pi}} \int_0^t \left\{ \alpha_{nn}(t-\tau) + \sum_{m=1, m\neq n}^{\infty} \alpha_{mn}(t-\tau) \right\} \frac{\dot{s}_n(\tau)}{\sqrt{t-\tau}} d\tau$$

$$+ \sqrt{\frac{\nu}{\pi}} \int_0^t \left\{ \beta_{nn}(t-\tau) + \sum_{m=1, m\neq n}^{\infty} \beta_{mn}(t-\tau) \right\} \frac{s_n(\tau)}{\sqrt{t-\tau}} d\tau = 0 \tag{4.2}$$

$(n = 1, 2, \ldots)$.

Equations (4.1) describe the perturbed rotational motion of a rigid body with a cavity containing a viscous fluid under the action of external moments, and equations (4.2) determine the fluid oscillation amplitudes $s_n(t)$.

Equations (4.2) contain the gyrostatic coefficients μ_n, $\boldsymbol{a}_n$, and $\boldsymbol{a}_n^*$ and the inertial coupling coefficients α_{mn}, α_{nn}, β_{mn}, and β_{nn}, which determine the influence of the fluid on the motion of the rigid body.

Let us write expressions for the inertial coupling coefficients for a cavity of arbitrary shape containing a viscous fluid:

$$\alpha_{mn} = \rho \int_S (V^*_m, [V_n \cos(2\omega_0, f)(t-\tau) + V_n \times f \sin(2\omega_0, f)(t-\tau)])dS,$$

$$\beta_{mn} = \rho \int_S (2\omega_0, f)(V^*_m, [V_n \sin(2\omega_0, f)(t-\tau) - V_n \times f \cos(2\omega_0, f)(t-\tau)])dS.$$

Here V_n is the velocity field of the ideal fluid, S is the surface of the fluid-containing cavity Q, and f is the inner normal to the surface of Q.

Let Q be a cylinder, that is,

$$Q = \{(R,\theta,z): \ 0 \leqslant R \leqslant 1, \ 0 \leqslant \theta \leqslant 2\pi, \ |z| \leqslant h\}.$$

In [106] the coefficients of inertial couplings of the body with the fluid in the case of a cylinder were calculated:

$$\begin{aligned} \alpha_{nn} &= \pi\rho T_n[A_n + B_n \cos 2\omega_0(t-\tau) + 2iC_n \sin 2\omega_0(t-\tau)], \\ \beta_{nn} &= 2\pi\rho\omega_0 T_n[B_n \sin 2\omega_0(t-\tau) - 2iC_n \cos 2\omega_0(t-\tau)], \\ \mu_n &= \pi\rho h T_n \chi_n C_n, \end{aligned} \tag{4.3}$$

where

$$A_n = h(1+k_l^2)(\chi_n^2 - 1), \quad B_n = k_l^2(\chi_n^2+1) + (\chi_n - 1)^2,$$

$$C_n = (1+k_l^2)\chi_n - 1, \quad T_n = \frac{\chi_n^2}{2\omega_0^2(\chi_n^2 - 1)},$$

$$\xi_n^2 = k_l^2(\chi_{lp}^2 - 1), \quad k_l = \frac{\pi}{2h}(2l+1),$$

$$n = (l,p), \quad l = 0,1,\ldots, \quad p = 1,2,\ldots,$$

χ_{lp} is the pth root of the equation $\xi J_1'(\xi) + \chi J_1(\xi) = 0$.

According to [106], the gyrostatic coefficients a_n have the form

$$a_n = \frac{4\chi_n(-1)^l}{\omega_0 \xi_n^4} \times [-i(\chi_n\xi_n^2 - (\chi_n - 1)^2)e_R + ((2+\chi_n)\xi_n^2 + (\chi_n^2 - 1)^2)e_\theta]. \tag{4.4}$$

In what follows, we need the complex conjugate gyrostatic coefficients a_n^*, which are

$$a_n^* = \frac{4\chi_n(-1)^l}{\omega_0 \xi_n^4}[i(\chi_n\xi_n^2 - (\chi_n - 1)^2)e_R + ((2+\chi_n)\xi_n^2 + (\chi_n^2 - 1)^2)e_\theta].$$

Solution of the system of equations of motion. We denote the Laplace image of an original $f(t)$ by $F(x)$; thus,

$$L(f(t)) = F(x) = \int_0^\infty e^{-xt} f(t)dt.$$

At the initial moment, the amplitude of the nth tone of oscillations of the fluid vanishes: $s_n(0) = 0$. Therefore,

$$L(\dot{s}_n(t)) = xS_n(x). \tag{4.5}$$

The Laplace transform of the convolution of two functions equals the product of the Laplace images of the functions being convolved; thus, using well-known formulas from [53], we obtain

$$\begin{aligned} &L\left(\int_0^t \dot{s}_n(\tau)\frac{d\tau}{\sqrt{t-\tau}}\right) = L(s_n)L\left(\frac{1}{\sqrt{t}}\right) = \sqrt{\pi x}S_n(x), \\ &L\left(\int_0^t s_n(\tau)\frac{\cos 2\omega_0(t-\tau)d\tau}{\sqrt{t-\tau}}\right) = \sqrt{\frac{\pi}{2}}S_n(x)\sqrt{\frac{\sqrt{x^2+4\omega_0^2}+x}{x^2+4\omega_0^2}}. \end{aligned} \tag{4.6}$$

Before applying the Laplace transform to system (4.1), (4.2), substitute the inertial couplings coefficients determined by (4.3) and (4.4) into this system. We obtain

$$\begin{aligned} &\mu_n\dot{s}_n(t) - i\lambda_n\mu_n s_n(t) + (a_n^*, \dot{\mathbf{\Omega}}(t)) + \rho\sqrt{\pi\nu}\left[\gamma_n\int_0^t \frac{\dot{s}_n(\tau)}{\sqrt{t-\tau}}d\tau\right. \\ &\left.+\delta_n\int_0^t \frac{\dot{s}_n(\tau)}{\sqrt{t-\tau}}\cos 2\omega_0(t-\tau)d\tau + 2i\varepsilon_n\int_0^t \frac{\dot{s}_n(\tau)}{\sqrt{t-\tau}}\sin 2\omega_0(t-\tau)d\tau\right] \\ &+2\rho\omega_0\sqrt{\pi\nu}\left[\delta_n\int_0^t \frac{s_n(\tau)}{\sqrt{t-\tau}}\sin 2\omega_0(t-\tau)d\tau\right. \\ &\left.-2\rho\varepsilon_n\int_0^t \frac{s_n(\tau)}{\sqrt{t-\tau}}\cos 2\omega_0(t-\tau)d\tau\right] = 0, \end{aligned} \tag{4.7}$$

where

$$\gamma_n = A_n T_n + \sum_{m=1,m\neq n}^{\infty} A_{mn} T_{mn}, \quad \delta_n = B_n T_n + \sum_{m=1,m\neq n}^{\infty} B_{mn} T_{mn},$$

$$\varepsilon_n = C_n T_n + \sum_{m=1,m\neq n}^{\infty} C_{mn} T_{mn}.$$

Let us apply the Laplace transform to system (4.7). In the space of Laplace images, the system of equations of motion takes the form

$$\begin{aligned} & S_n(x)\Bigg\{\mu_n x - i\lambda_n\mu_n + \pi\rho\sqrt{\nu}\gamma_n\sqrt{x} \\ & + \pi\rho\sqrt{\frac{\nu}{2}}\left(\delta_n x\sqrt{\frac{\sqrt{x^2+4\omega_0^2}+x}{x^2+4\omega_0^2}} + 2i\varepsilon_n x\sqrt{\frac{\sqrt{x^2+4\omega_0^2}-x}{x^2+4\omega_0^2}}\right) \\ & + \sqrt{2\nu}\pi\rho\omega_0\left(\delta_n\sqrt{\frac{\sqrt{x^2+4\omega_0^2}-x}{x^2+4\omega_0^2}} - 2i\varepsilon_n\sqrt{\frac{\sqrt{x^2+4\omega_0^2}+x}{x^2+4\omega_0^2}}\right)\Bigg\} \\ & = -(a_n^*, L(\dot{\boldsymbol{\Omega}}(t))) \quad (n = 1, 2, \dots). \end{aligned} \tag{4.8}$$

System (4.8) linearized at small x by discarding the terms containing x^2 and higher powers of x is

$$\begin{aligned} & S_n(x)\Bigg\{\mu_n x - i\lambda_n\mu_n + \pi\rho\sqrt{\nu}\gamma_n\sqrt{x} + \pi\rho\sqrt{\frac{\nu}{\omega_0}}(\delta_n + 2i\varepsilon_n)x \\ & + \sqrt{2\nu}\pi\rho\omega_0\left(\delta_n\frac{4\omega_0 - x}{4\omega_0\sqrt{2\omega_0}} - 2i\varepsilon_n\frac{4\omega_0 + x}{4\omega_0\sqrt{2\omega_0}}\right)\Bigg\} \\ & = -(a_n^*, L(\dot{\boldsymbol{\Omega}}(t))) \quad (n = 1, 2, \dots). \end{aligned}$$

Thus,

$$S_n(x) = -\frac{(a_n^*, L(\dot{\boldsymbol{\Omega}}(t)))}{D_n + \varepsilon_n\sqrt{x} + F_n x}, \tag{4.9}$$

where

$$D_n = -i\lambda_n\mu_n + \pi\rho\sqrt{\nu\omega_0}(\delta_n - 2i\varepsilon_n), \quad \varepsilon_n = \pi\rho\sqrt{\nu}\gamma_n,$$

$$F_n = \mu_n + \frac{1}{4}\pi\rho\sqrt{\frac{\nu}{\omega_0}}(\delta_n + 2i\varepsilon_n).$$

Let us further simplify formula (4.9) at small x:

$$S_n(x) = -\frac{(a_n^*, L(\dot{\mathbf{\Omega}}(t)))}{D_n + \varepsilon_n\sqrt{x}}. \tag{4.10}$$

We use this formula (which is valid only at small x) to determine the nth tone of oscillations of the viscous fluid at large t.

Performing the inverse Laplace transform we obtain

$$s_n(t) = L^{-1}[S_n(x)] = L^{-1}\left[-\frac{(a_n^*, L(\dot{\mathbf{\Omega}}(t)))}{D_n + \varepsilon_n\sqrt{x}}\right].$$

Thus, we found the oscillation amplitudes of a viscous fluid in a cavity of a rotating body.

Consider various cases of the perturbed motion $\mathbf{\Omega}(t)$ relative to steady rotation.

1. Let $L\left(\dot{\Omega}(t)\right) = \boldsymbol{l}/\sqrt{x}$, where $\boldsymbol{l}$ is a constant vector.

Applying the inverse Laplace transform to (4.10) at large t, we obtain

$$s_n(t) = -\frac{(a_n^*, \boldsymbol{l})}{\varepsilon_n} e^{(\frac{D_n}{\varepsilon_n})^2 t} \operatorname{erfc}\left(\frac{D_n}{\varepsilon_n}\sqrt{t}\right).$$

In this case, we can also find the vector of perturbed motion:

$$\mathbf{\Omega}(t) = 2\boldsymbol{l}\sqrt{\frac{t}{\pi}}.$$

2. Let $L(\dot{\mathbf{\Omega}}(t)) = \boldsymbol{l}e^{-k\sqrt{x}}$ $(k \geqslant 0)$.

Then

$$s_n(t) = \frac{(a_n^*, \boldsymbol{l})}{\varepsilon_n}\left\{\frac{1}{\sqrt{\pi t}}e^{-\frac{k^2}{4t}} - \frac{D_n}{\varepsilon_n}e^{\frac{D_n}{\varepsilon_n}k}e^{(\frac{D_n}{\varepsilon_n})^2 t}\operatorname{erfc}\left(\frac{D_n}{\varepsilon_n}\sqrt{t} + \frac{k}{2\sqrt{t}}\right)\right\}.$$

In this case, we have

$$\dot{\mathbf{\Omega}}(t) = \boldsymbol{l}\frac{k}{2\sqrt{\pi t^3}}e^{-\frac{k^2}{4t}}.$$

3. Let $L(\dot{\mathbf{\Omega}}(t)) = \boldsymbol{l}e^{-k\sqrt{x}}/\sqrt{x}$ $(k \geqslant 0)$. Then

$$s_n(t) = -\frac{(a_n^*, \boldsymbol{l})}{\varepsilon_n}e^{\frac{D_n}{\varepsilon_n}k}e^{\frac{D_n}{\varepsilon_n}t}\operatorname{erfc}\left(\frac{D_n}{\varepsilon_n}\sqrt{t} + \frac{k}{2\sqrt{t}}\right).$$

In this case, we have

$$\dot{\mathbf{\Omega}}(t) = \boldsymbol{l}\frac{1}{\sqrt{\pi t}}e^{-\frac{k^2}{4t}}.$$

Thus, we have found the velocity field for a viscous fluid in a rotating cylinder. Knowing this field, we can determine the tangential stress tensions vector acting from the side of the fluid on the rigid body and, thereby, the moment of friction forces.

5 INTERNAL FRICTION MOMENT IN A FLUID-FILLED GYROSCOPE

This section considers weakly excited motions of a rapidly rotating vessel completely filled with a fluid. The vessel is assumed to be an absolutely rigid solid body, and the fluid is assumed to be viscous. The equations of motion are linearized in a neighborhood of uniform rotation. The moment of internal friction forces acting from the side of the fluid on the vessel shell. We obtain expressions for the internal friction moment under the action of perturbations of various types from the side of the fluid on the vessel.

Consider a rigid body with a cavity filled with a viscous incompressible fluid of density ρ and kinematic viscosity ν.

We assume that the unperturbed motion is the steady rotation at an angular velocity $\boldsymbol{\omega}_0 = \omega_0 \boldsymbol{k}$, where $\boldsymbol{k}$ is the basis vector along the Ox_3 axis of a coordinate system $Ox_1x_2x_3$ attached to the body.

We denote the vector of small oscillations of the gyroscope relative to the steady rotation by $\boldsymbol{\Omega}(t)$, so that the angular velocity of the body can be represented in the form $\boldsymbol{\omega} = \boldsymbol{\omega}_0 + \boldsymbol{\Omega}(t)$, and $\boldsymbol{\Omega}(t)$ is first-order small in comparison with $\boldsymbol{\omega}_0$.

Let I_0 be the inertia tensor of the rigid body without fluid, and let I be the inertia tensor of the hardened fluid; by $\boldsymbol{M}$ we denote the moment of the external forces acting on the body (without fluid). The coefficients characterizing the inertial relation of the body and the fluid are expressed in terms of solutions of boundary value problems depending only on the geometry of the cavity as

$$\mu_n = \rho \int\limits_Q (\boldsymbol{V}_n, \boldsymbol{V}_n^*) dQ, \quad \boldsymbol{a}_n = \rho \int\limits_Q \boldsymbol{r} \times \boldsymbol{V}_n \, dQ, \quad \lambda_n = \frac{2\omega_0}{\chi_n}. \tag{5.1}$$

Here Q is the volume of the cavity, the χ_n and the $\boldsymbol{V}_n$ are the eigenfrequencies and the natural forms of oscillations of the ideal fluid in the cavity of the body rotating at a constant angular velocity $\boldsymbol{\omega}_0$, and $\boldsymbol{r}$ is radius vector.

The velocity vector of the viscous fluid can be represented as

$$\boldsymbol{V} = \sum_{n=1}^{\infty} S_n(t) \boldsymbol{V}_n(x_1, x_2, x_3), \tag{5.2}$$

where $S_n(t)$ has the meaning of the amplitude of the nth tone of fluid oscillations and $\boldsymbol{V}_n$ is the velocity field of the ideal fluid found in [106].

The equations of the weakly perturbed motion of a body with a fluid have the form

$$(I + I^0)\dot{\boldsymbol{\Omega}} + \boldsymbol{\Omega} \times (I + I^0)\boldsymbol{\omega}_0 + \boldsymbol{\omega}_0 \times (I + I^0)\boldsymbol{\Omega} + \sum_{n=1}^{\infty} [\boldsymbol{a}_n \dot{S}_n + (\boldsymbol{\omega}_0 \times \times \boldsymbol{a}_n) S_n] = \boldsymbol{M}, \tag{5.3}$$

$$\mu_n \dot{S}_n - i\lambda_n \mu_n S_n + (a_n^*, \dot{\Omega}) + \sqrt{\frac{\nu}{\pi}} \int_0^t \left\{ \alpha_{nn}(t-\tau) + \sum_{m=1, m\neq n}^{\infty} \alpha_{mn}(t-\tau) \right\} \frac{\dot{S}_n(\tau)}{\sqrt{t-\tau}} d\tau$$

$$+ \sqrt{\frac{\nu}{\pi}} \int_0^t \left\{ \beta_{nn}(t-\tau) + \sum_{m=1, m\neq n}^{\infty} \beta_{mn}(t-\tau) \right\} \frac{S_n(\tau)}{\sqrt{t-\tau}} d\tau = 0 \tag{5.4}$$

$(n = 1, 2, \dots)$.

Here equation (5.3) describes the perturbed motion of a rigid body with a fluid under the action of external moments and system (5.4) determines the oscillation amplitudes $S_n(t)$ of the viscous fluid.

The system of equations (5.4) contains other coefficients characterizing the influence of the fluid on the motion of the rigid body, namely, α_{mn}, α_{nn}, β_{mn}, and β_{nn}.

It follows from equation (5.3) that the moment of the forces acting on the body from the side of the fluid in the relative motion equals

$$M = -\sum_{n=1}^{\infty} [a_n \dot{S}_n + (\omega_0 \times a_n) S_n], \tag{5.5}$$

where $S_n(t)$ is the generalized coordinate characterizing the nth tone of fluid oscillations.

To determine the moment of the friction forces acting from the fluid on the shell, we apply the Laplace transform.

We denote the Laplace image of the an $f(t)$ by $F(x)$; i.e., we set

$$L(f(t)) = F(x) = \int_0^{\infty} e^{-xt} f(t) dt. \tag{5.6}$$

By the conditions of the problem, the amplitude of the nth tone of fluid oscillations vanishes at the initial moment time, that is, $S_n(+0) = 0$.

Therefore,

$$L(\dot{s}_n(t)) = x S_n(x). \tag{5.7}$$

The moments of the forces acting on the body from the side of the fluid in the absence of viscosity are determined by relation (5.5) with $\nu = 0$, i.e., by

$$L(M|_{\nu=0}) = -\sum_{n=1}^{\infty} [a_n x + \omega_0 \times a_n] S_n(x)|_{\nu=0}. \tag{5.8}$$

In section 4 of this chapter, we found the Laplace transform of the generalized coordinate $S_n(t)$ characterizing the nth tone of fluid oscillations in a cylindrical cavity:

$$S_n(x)|_{\nu=0} = -\frac{(a_n^*, L(\dot{\Omega}(t)))}{\mu_n(x - i\lambda_n)}. \tag{5.9}$$

Here the asterisk denotes complex conjugation.

Now, taking into account relation (5.9) and using (5.7) and (5.8), we obtain the Laplace transform of the friction moment:

$$L(M_{TP}) = L(M) - L(M|_{\nu=0}),$$
$$L(M_{TP}) = \sum_{n=1}^{\infty} [a_n x + \omega_0 \times a_n] \times \left[\frac{(a_n^*, L(\dot{\Omega}(t)))}{D_n + E_n\sqrt{x} + F_n x)} - \frac{(a_n^*, L(\dot{\Omega}(t)))}{\mu_n(x - i\lambda_n)} \right]. \tag{5.10}$$

Here

$$D_n = -i\lambda_n\mu_n + \pi\rho\sqrt{\nu\omega_0}(\delta_n - 2i\varepsilon_n), \quad E_n = \pi\rho\sqrt{\nu}\gamma_n,$$
$$F_n = \mu_n + \frac{1}{4}\pi\rho\sqrt{\frac{\nu}{\omega_0}}(\delta_n + 2i\varepsilon_n), \quad \mu_n = \pi\rho h\chi_n T_n C_n, \tag{5.11}$$

where the following notation is used:

$$\gamma_n = A_n T_n + \sum_{m=1, m\neq n}^{\infty} A_{mn} T_{mn}, \quad \delta_n = B_n T_n + \sum_{m=1, m\neq n}^{\infty} B_{mn} T_{mn},$$
$$\varepsilon_n = C_n T_n + \sum_{m=1, m\neq n}^{\infty} C_{mn} T_{mn},$$

where

$$A_n = h(1 + k_l^2)(\chi_n^2 - 1), \quad B_n = k_l^2(\chi_n^2 + 1) + (\chi_n - 1)^2,$$
$$C_n = (1 + k_l^2)\chi_n - 1, \quad T_n = \frac{\chi_n^2}{2\omega_0^2(\chi_n^2 - 1)}, \quad \lambda_n = \frac{2\omega_0}{\chi_n},$$

and the χ_n are the eigenfrequencies of the fluid oscillations in the cavity of the body rotating at a constant velocity ω_0.

Relation (5.10) determines the Laplace transform of the friction moment.

For small ν, expression (5.10) can be simplified. Expanding (5.10) in a Taylor series in powers of $\sqrt{\nu}$ and retaining only linear terms, we obtain

$$L(M_{TP}) = -\pi\rho\sqrt{\nu}\sum_{n=1}^{\infty} \frac{(a_n x + \omega_0 \times a_n)}{\mu_n^2(x - i\lambda_n)^2}(a_n^*, L(\dot{\Omega}))$$
$$\times \left[\sqrt{\omega_0}(\delta_n - 2i\varepsilon_n) + \gamma_n\sqrt{x} + \frac{1}{4\sqrt{\omega_0}}(\delta_n + 2i\varepsilon_n)x \right]. \tag{5.12}$$

At small x, expression (5.12) transforms into

$$L(M_{TP}) = \pi\rho\sqrt{\nu}\sum_{n=1}^{\infty} \frac{(\omega_0 \times a_n)(a_n^*, L(\dot{\Omega}))}{\lambda_n^2\mu_n^2} \times [\sqrt{\omega_0}(\delta_n - 2i\varepsilon_n) + \gamma_n\sqrt{x}]. \tag{5.13}$$

This and the linearity of the inverse Laplace transform L^{-1} imply the following expression for the friction moment:

$$M_{TP} = \pi\rho\sqrt{\nu}\sum_{n=1}^{\infty}\frac{\omega_0 \times a_n}{\lambda_n^2\mu_n^2} \times (a_n^*, [\sqrt{\omega_0}(\delta_n - 2i\varepsilon_n)\dot{\boldsymbol{\Omega}} + L^{-1}(\sqrt{x}L(\dot{\boldsymbol{\Omega}}))\gamma_n]). \quad (5.14)$$

Consider various cases of perturbations acting on a rotating rotor.

Let $\boldsymbol{\Omega}(t) = 2l\sqrt{t/\pi}$, where l is a constant vector.

Then $\dot{\boldsymbol{\Omega}} = l/\sqrt{\pi t}$; therefore, $L(\dot{\boldsymbol{\Omega}}) = l/\sqrt{x}$. Substituting these expressions into (5.14), we find the following expression for the internal friction moment:

$$M_{TP} = \pi\rho\sqrt{\nu}\sum_{n=1}^{\infty}\frac{\omega_0 \times a_n}{\lambda_n^2\mu_n^2}\left(a_n^*, l\left[\sqrt{\frac{\omega_0}{\pi t}}(\delta_n - 2i\varepsilon_n) + \gamma_n\delta(t)\right]\right),$$

where $\delta(t)$ is the Dirac delta-function.

Suppose that the acceleration of the perturbed motion is given and equals

$$\dot{\boldsymbol{\Omega}}(t) = l\,\mathrm{erfc}\frac{a}{2\sqrt{t}},$$

where $a = \mathrm{const}$, $a > 0$, and $\mathrm{erfc}\,x = \frac{2}{\sqrt{\pi}}\int_x^{\infty} e^{-u^2}du$.

Then

$$L(\dot{\boldsymbol{\Omega}}) = \frac{e^{-a\sqrt{x}}}{x}l, \quad L^{-1}\left(\frac{e^{-a\sqrt{x}}}{\sqrt{x}}l\right) = \frac{1}{\sqrt{\pi t}}e^{-\frac{a^2}{4t}}l.$$

Substituting these relations into (5.14), we obtain

$$M_{TP} = \pi\rho\sqrt{\nu}\sum_{n=1}^{\infty}\frac{\omega_0 \times a_n}{\lambda_n^2\mu_n^2}\left(a_n^*, l\left[\mathrm{erfc}\frac{a}{2\sqrt{t}}\sqrt{\omega_0}(\delta_n - 2i\varepsilon_n) + \gamma_n\frac{1}{\sqrt{\pi t}}e^{-\frac{a^2}{4t}}\right]\right).$$

If the acceleration of the perturbed motion has the form

$$\dot{\boldsymbol{\Omega}}(t) = \frac{l}{\sqrt{\pi t}}e^{-\frac{a^2}{4t}},$$

then

$$L(\dot{\boldsymbol{\Omega}}) = \frac{e^{-a\sqrt{x}}}{\sqrt{x}}l, \quad L^{-1}(le^{-a\sqrt{x}}) = \frac{al}{2\sqrt{\pi}t^{3/2}}e^{-\frac{a^2}{4t}}.$$

Substituting these relations into (5.14), we obtain the friction moment

$$M_{TP} = \pi\rho\sqrt{\nu}\sum_{n=1}^{\infty}\frac{\omega_0 \times a_n}{\lambda_n^2\mu_n^2} \times \left(a_n^*, l\left[\sqrt{\omega_0}(\delta_n - 2i\varepsilon_n)\frac{1}{\sqrt{\pi t}}e^{-\frac{a^2}{4t}} + \frac{\gamma_n a}{2\sqrt{\pi}t^{3/2}}e^{-\frac{a^2}{4t}}\right]\right).$$

Thus, we have obtained an explicit expression for the moment of the internal friction forces acting on the shell of a rapidly rotating vessel completely filled with a viscous fluid. This result can be used both for calculating constructive-technological parameters of engineering objects and for comparing conclusions of boundary layer theory with results of the corresponding experimental studies.

6 STABILITY OF A FLUID-FILLED GYROSCOPE

This section studies the stability of the steady rotation of a symmetric body with a viscous fluid on the basis of integro-differential equations with coefficients determined by solving hydrodynamic boundary value problems for an ideal fluid which depend on the geometry of the cavity.

We solve the problem of the stability of the rotation of a body about the axis corresponding to the greatest moment of inertia and the instability of rotation about the axis corresponding to the least moment of inertia by the perturbation method.

A similar problem was considered in [34, 148, 299] for a body with an ideal fluid and in paper [58] for a body with a viscous fluid.

Consider the motion of a dynamically symmetric body with an axially symmetric cavity completely filled with a low-viscosity incompressible fluid perturbed relative to the steady rotation. The angular velocity of the body can be represented in the form

$$\boldsymbol{\omega} = \boldsymbol{\omega}_0 + \boldsymbol{\Omega}(t) = \omega_0 \boldsymbol{k} + \boldsymbol{\Omega}(t).$$

Here $\boldsymbol{\omega}_0 = \omega_0 \boldsymbol{k}$ is the angular velocity of the steady rotation of the body directed along the basis vector $\boldsymbol{k}$ of the Ox_3 axis of the body coordinate system $Ox_1x_2x_3$; $\boldsymbol{\Omega} = (\Omega_1, \Omega_2, 0)$ is the angular velocity of the body in the perturbed motion, which is first-order small in comparison with $\boldsymbol{\omega}_0$.

The equations of perturbed motion can be written in the form [105]

$$A\dot{\Omega} + i(C - A)\omega_0 \Omega + 2\rho \sum_{n=1}^{\infty} a_n(\dot{s}_n - i\omega_0 s_n) = M,$$

$$\mu_n^2 \left\{ \dot{s}_n - i\lambda_n s_n + \sqrt{\frac{\nu}{\pi}} \int_0^t \frac{\dot{s}_n(\tau)\alpha_n(t-\tau) + s_n(\tau)\beta_n(t-\tau)}{\sqrt{t-\tau}} d\tau \right\} + a_n^* \dot{\Omega} \tag{6.1}$$

$$= -\sqrt{\frac{\nu}{\pi}} \sum_{m=1, n\neq m}^{\infty} \int_0^t \frac{\dot{s}_n(\tau)\alpha_{mn}(t-\tau) + s_n(\tau)\beta_{mn}(t-\tau)}{\sqrt{t-\tau}} d\tau.$$

Here $\Omega = \Omega_1 - i\Omega_2$, $M = M_1 - iM_2$, and A and C are the inertia moments of the body–fluid system relative to the axis of symmetry and the transverse axis.

Since the cross coefficients α_{mn} and β_{mn} of inertial couplings weakly affect the dynamics of a rotor with a viscous fluid, we can neglect their influence and retain only the principal terms with $m = n$.

Let us introduce a function $\phi_n(x, R)$ satisfying the two-dimensional boundary value problem

$$\begin{gathered}\frac{\partial^2\phi_n}{\partial R^2}+\frac{1}{R}\frac{\partial\phi_n}{\partial R}-\frac{\phi_n}{R^2}+(1-\chi_n^2)\frac{\partial^2\phi_n}{\partial x^2}=0,\\ \left(\frac{\partial\phi_n}{\partial R}+\chi_n\frac{\phi_n}{R}\right)n_R+(1-\chi_n^2)\frac{\partial\phi_n}{\partial x}n_x=0.\end{gathered}\tag{6.2}$$

The first equation must be satisfied in the meridional section G of the cavity in the plane of cylindrical coordinates (R, x) and the second, on the boundary of this section, which has normal $\boldsymbol{n} = (n_R, n_x)$; $\chi_n = 2\omega_0/\lambda_n$. The coefficients in the equations are expressed in terms of the function ϕ_n as

$$\begin{gathered}a_n=-\frac{\pi\rho\chi_n}{2\omega_0}\int\limits_G\left[R\frac{\partial\phi_n}{\partial x}+\frac{x}{\chi_n-1}\left(\frac{\partial\phi_n}{\partial R}+\frac{\phi_n}{R}\right)\right]R\,dS,\\ \mu_n^2=\frac{\pi\rho\chi_n^2}{2\omega_0^2}\times\int\limits_G\left\{\left(\frac{\partial\phi_n}{\partial x}\right)^2+\frac{\chi_n^2+1}{(\chi_n^2-1)^2}\left[\left(\frac{\partial\phi_n}{\partial R}\right)^2+\frac{\phi_n^2}{R^2}\right]+\frac{4\chi_n}{(\chi_n^2-1)^2}\frac{\partial\phi_n}{\partial R}\frac{\phi_n}{R}\right\}R\,dS,\end{gathered}\tag{6.3}$$

$$\begin{gathered}\alpha_n(t-\tau)=A_n+B_n\cos 2\omega_0(t-\tau)+2iC_n\sin 2\omega_0(t-\tau),\\ \beta_n(t-\tau)=2\omega_0 B_n\sin 2\omega_0(t-\tau)-4i\omega_0 C_n\cos 2\omega_0(t-\tau);\end{gathered}\tag{6.4}$$

for a cylindrical cavity of radius $r_0 = 1$ and height h, the following coefficients were found in [105]:

$$\begin{gathered}A_n=T_nh(1+k_l^2)(\chi_n^2-1),\quad B_n=T_n[k_l^2(\chi_n^2+1)+(\chi_n-1)^2],\\ C_n=T_n(1+k_l^2)\chi_n-1,\quad T_n=\frac{\pi\rho\chi_n^2}{2\omega_0^2(\chi_n^2-1)}.\end{gathered}\tag{6.5}$$

Let us write the characteristic equation for system (6.1) by setting

$$\begin{gathered}A\dot{\Omega}+i(C-A)\omega_0\Omega+2\rho\sum_{n=1}^{\infty}a_n(\dot{s}_n-i\omega_0 s_n)=0,\\ \mu_n^2\left\{\dot{s}_n-i\lambda_n s_n+\sqrt{\frac{\nu}{\pi}}\int\limits_0^t\frac{\dot{s}_n(\tau)\alpha_n(t-\tau)+s_n(\tau)\beta_n(t-\tau)}{\sqrt{t-\tau}}d\tau\right\}+a_n\dot{\Omega}=0.\end{gathered}\tag{6.6}$$

Applying the Laplace transform to equations (6.6), we obtain

$$[Ap + i(C - A)\omega_0]\hat{\Omega} + 2\rho \sum_{n=1}^{\infty} a_n(p - i\omega_0)\hat{S}_n = 0,$$

$$\mu_n^2 \left[p - i\lambda_n + \sqrt{\frac{1}{2}}\sqrt{\nu}B_n(p + 2\omega_0)\left(\frac{1}{\sqrt{p + i2\omega_0}} + \frac{1}{\sqrt{p - i2\omega_0}}\right) \right. \tag{6.7}$$

$$\left. -\sqrt{\nu}C_n(p - 2\omega_0)\left(\frac{1}{\sqrt{p + i2\omega_0}} - \frac{1}{\sqrt{p - i2\omega_0}}\right) + \frac{A_n}{p}\sqrt{\nu}\right]\hat{S}_n + a_n p\hat{\Omega} = 0.$$

The characteristic equation for system (6.1) has the form

$$Ap + i(C - A)\omega_0 - p(p - i\omega_0)\sum_{n=1}^{\infty} \frac{E_n}{\psi_n(p)} = 0, \tag{6.8}$$

where $E_n = 2\rho a_n^2/\mu_n^2$,

$$\psi_n(p) = p - i\lambda_n + \frac{A_n}{p}\sqrt{\nu} + \frac{1}{2}\sqrt{\nu}B_n(p + 2\omega_0) \times \left(\frac{1}{\sqrt{p + i2\omega_0}} + \frac{1}{\sqrt{p - i2\omega_0}}\right)$$

$$- \sqrt{\nu}C_n(p - 2\omega_0)\left(\frac{1}{\sqrt{p + i2\omega_0}} - \frac{1}{\sqrt{p - i2\omega_0}}\right). \tag{6.9}$$

Let us calculate the roots of the function $\psi_n(p)$ by the perturbation method, retaining only the terms linear in the small parameter $\sqrt{\nu}$. Suppose that

$$p_n = i\lambda_n + \sqrt{\nu}\delta_n, \quad \sqrt{\nu} \ll 1, \quad \psi_n(p_n) = 0. \tag{6.10}$$

Then

$$\begin{aligned}\delta_n = & \frac{1}{\sqrt{2\omega_0 + \lambda_n}}(B_n(2\omega_0 - \lambda_n) + C_n(2\omega_0 + \lambda_n)) \\ & + i\left[\frac{A_n}{\lambda_n} + \frac{1}{\sqrt{2\omega_0 + \lambda_n}}(C_n(2\omega_0 + \lambda_n) - B_n(2\omega_0 - \lambda_n)) \right. \\ & \left. + \frac{1}{\sqrt{2\omega_0 - \lambda_n}}(B_n(2\omega_0 + \lambda_n) + C_n(2\omega_0 - \lambda_n))\right].\end{aligned} \tag{6.11}$$

Taking into account the fact that stability is lost at frequencies close to partial oscillation frequencies of the fluid, we expand the meromorphic function $1/\psi_n(p)$ in a Laurent series and consider only the terms in neighborhoods of the poles, i.e., the zeros of the function $\psi_n(p)$. Then the characteristic equation (6.8) takes the form

$$Ap + i(C - A)\omega_0 - p(p - i\omega_0)\sum_{n=1}^{\infty} \frac{E_n}{(p - p_n)\psi_n'(p_n)} = 0. \tag{6.12}$$

Let us calculate the derivative of $\psi_n(p_n)$, retaining only the terms of order $\sqrt{\nu}$:

$$\psi'_n(i\lambda_n) = 1 + \sqrt{\nu}\tilde{\alpha}_n + i\sqrt{\nu}\tilde{\beta}_n, \tag{6.13}$$

where

$$\begin{aligned}\tilde{\alpha}_n &= \frac{A_n}{\lambda_n^2} + \frac{B_n}{4\sqrt{2}}\{2(\gamma^+ + \gamma^-) - 2\omega_0(\gamma^{+3} + \gamma^{-3}) + \lambda_n(\gamma^{+3} - \gamma^{-3})\} \\ &\quad - \frac{C_n}{2\sqrt{2}}\{2(\gamma^+ + \gamma^-) + 2\omega_0(\gamma^{+3} - \gamma^{-3}) + \lambda_n(\gamma^{+3} + \gamma^{-3})\}, \\ \tilde{\beta}_n &= \frac{B_n}{4\sqrt{2}}\{2(\gamma^- - \gamma^+) - 2\omega_0(\gamma^{+3} - \gamma^{-3}) - \lambda_n(\gamma^{+3} + \gamma^{-3})\} \\ &\quad - \frac{C_n}{2\sqrt{2}}\{2(\gamma^- - \gamma^+) + 2\omega_0(\gamma^{+3} + \gamma^{-3}) - \lambda_n(\gamma^{+3} - \gamma^{-3})\}.\end{aligned} \tag{6.14}$$

where $\gamma^{\pm} = 1/\sqrt{2\omega_0 \pm \lambda_n}$.

Equation (6.12) with (6.13) and (6.14) taken into account has the form

$$Ap + i(C - A)\omega_0 - p(p - i\omega_0)\sum_{n=1}^{\infty} \frac{L_{1n} - iL_{2n}}{p - p_n} = 0. \tag{6.15}$$

Here

$$L_{1n} = E_n(1 - \tilde{\alpha}_n\sqrt{\nu}), \quad L_{2n} = \tilde{\beta}_n E_n \sqrt{\nu}. \tag{6.16}$$

Equation (6.15) is a generalization of the characteristic equation paper [57] to the case of a low-viscosity fluid.

In the absence of viscosity ($\sqrt{\nu} = 0$), in the case of an ellipsoidal cavity, equation (6.15) coincides with the characteristic equations obtained in [148, 299] (it suffices to set gravity to zero in [148, 299]) and in the case of an arbitrary cavity of revolution, with an equation in [57].

Let us set $\nu = 0$ and make the change $p = i\eta$. Then the stability of the steady rotation requires that all roots η be real. As a numerical analysis performed in [57] shows, in the first approximation, we can leave only one principal term (corresponding to $n = 1$) in the infinite sum. The equations of the boundary of the stability domain satisfy the relations

$$\Delta = C - A = -E_1 - (A - 2E_1)\frac{2}{\chi_1} \pm \frac{2}{\chi_1}\sqrt{(A - E_1)E_1^2 - (\chi_1 - 2)}; \tag{6.17}$$

the domain of instability is between the curves determined by the positive and negative values of the radical. It follows from expression (6.17) that the rotation of the body is stable at $C > A$, i.e., when the body rotates about the axis corresponding to the greatest moment of inertia.

Let us prove that the free rotation of a body with an axially symmetric cavity about the axis corresponding to the least moment of inertia ($\Delta < 0$) is always unstable.

Considering the limit case $\omega_0 \to 0$ of equation (6.15) with $\nu = 0$, we can show that the quantity

$$A' = A - \sum_{n=1}^{\infty} E_n > 0 \tag{6.18}$$

equals the moment of inertia of the equivalent rigid body described by Zhukovskii in [332].

The infinite sum $\sum_n E_n$ being bounded by the value A of the moment of inertia suggests that a reduction of the infinite system (6.15) may be possible. Let us write the characteristic equation (6.15) in which only a finite number N of terms in the infinite series are left:

$$A\eta + \Delta - \eta(\eta - 1) \sum_{n=1}^{N} \frac{E_n}{\eta - b_n} = 0, \tag{6.19}$$

where $b_n = \lambda_n/\omega_0$. In a small neighborhood of the point $\eta = b_n$, equation (6.19) is equivalent to the quadratic equation

$$\eta^2(A - E_n) + \eta(\Delta + E_n - Ab_n) - \Delta b_n = 0. \tag{6.20}$$

This quadratic equation has complex roots (the rotation of the body is unstable) if its discriminant is negative, i.e.,

$$D = (\Delta + E_n - Ab_n)^2 + 4\Delta b_n(A - E_n) < 0. \tag{6.21}$$

The relation $D = 0$, which is itself a quadratic equation with respect to b_n, provides the intervals of dimensionless eigenfrequencies corresponding to the unstable rotation. These frequencies lie between the roots b_n^0 of the equation $D = 0$:

$$b_n^0 = \frac{1}{A^2}\Big[-\Delta A + E_n A + 2\Delta E_n \pm\sqrt{(\Delta A - E_n A - 2\Delta E_n)^2 - A^2(\Delta^2 + E_n^2 + 2\Delta E_n)}\Big]. \tag{6.22}$$

The radicand is always positive:

$$(\Delta A - E_n A - 2\Delta E_n)^2 - A^2(\Delta^2 + E_n^2 + 2\Delta E_n) = 4\Delta A E_n(E_n - A) > 0,$$

i.e., $\Delta < 0$ and $A > \sum_{k=1}^{\infty} E_k > E_n$. Since $\Delta < 0$, we have $b_n^0 \to 1 - C/A < 1$ as $E_n \to 0$, and since the spectrum of the eigenfrequencies is dense in the domain $|b_n| < 1$, we can always choose a real value b_n in the interval determined by (6.22) with the corresponding coefficient E_n (which may be fairly small) in whose neighborhood $\Delta < 0$, so that equation (6.20) has complex roots, which means the instability of the body rotation about the axis corresponding to the least moment of inertia. It is easy to estimate the imaginary part of the root (6.20), which determines the order of the intensity of the stability loss in the steady rotation (i.e., the increment of rotation axis) at the given frequency b_n:

$$\operatorname{Im}\eta = \frac{\sqrt{-D}}{A}.$$

Substituting the mean frequency from the interval (6.22) into the expression for D from (6.21), we obtain

$$\operatorname{Im}\eta = \sqrt{\left(1 - \frac{E_n}{A}\right)\frac{E_n}{A}\left(1 - \frac{C}{A}\right)\frac{C}{A}}.$$

This gives the characteristic time of the stability loss by the body rotating at an angular velocity ω_0:

$$T = \frac{1}{\omega_0}\left[\left(1 - \frac{E_n}{A}\right)\frac{E_n}{A}\left(1 - \frac{C}{A}\right)\frac{C}{A}\right]^{1/2}.$$

In accordance with the above considerations, we leave one term in the infinite sum in equation (6.15) and make the changes

$$p = i\eta, \quad p_n = i\lambda_n + \sqrt{\nu}\delta_n.$$

Let us write the characteristic equation as

$$A\eta + (C - A)\omega_0 - \eta(\eta - \omega_0)\frac{L_{11} - iL_{21}}{(\eta - \lambda_1) - i\sqrt{\nu}\delta_1} = 0, \tag{6.23}$$

or, in a more convenient form, as

$$A\eta + (C - A)\omega_0 - \frac{\eta(\eta - \omega_0)}{\eta - \lambda_1}E_1\left\{1 + \sqrt{\nu}\left(-\alpha_1 - i\beta_1 + i\frac{\delta_1}{\eta - \lambda_1}\right)\right\} = 0. \tag{6.24}$$

Let

$$\Delta = -\alpha_1 - i\beta_1 + i\frac{\delta_1}{\eta - \lambda_1}.$$

We seek the viscousity correction to the root by the method of perturbation theory. Let η^0 be a root of the characteristic equation for an ideal fluid:

$$A\eta^0(\eta^0 - \lambda_1) - \eta^0(\eta^0 - \omega_0)E_1 + (C - A)\omega_0(\eta^0 - \lambda_1) = 0. \tag{6.25}$$

The characteristic equation for for a viscous fluid has the form

$$A\eta(\eta - \lambda_1) + (C - A)\omega_0(\eta - \lambda_1) - \eta(\eta - \omega_0)E_1[1 + \sqrt{\nu}\Delta] = 0. \tag{6.26}$$

We shall seek the correction to the root in the form

$$\eta = \eta^0 + \sqrt{\nu}\Delta^*. \tag{6.27}$$

Substituting (6.27) into (6.26) and retaining only terms linear in $\sqrt{\nu}$, we obtain the correction Δ^*:

$$\Delta^* = \frac{\eta^0(\eta^0 - \omega_0)E_1\Delta}{A(2\eta^0 - \lambda_1) - E_1(2\eta^0 - \omega_0) + (C - A)\omega_0}. \tag{6.28}$$

To ensure the stability of a rotor with a viscous fluid, we write an explicit expression for p:

$$p = i\eta = i\eta^0 + \sqrt{\nu}i\Delta^* = i\eta^0 + \sqrt{\nu}i\Delta\Xi,$$

where

$$\Xi = \frac{\eta^0(\eta^0 - \omega_0)E_1}{A(2\eta^0 - \lambda_1) - E_1(2\eta^0 - \omega_0) + (C - A)\omega_0}. \tag{6.29}$$

Consider $i\Delta$ separately:

$$i\Delta = -i\alpha_1 + \beta_1 - \frac{\delta_{11}}{\eta^0 - \lambda_1} - i\frac{\delta_{12}}{\eta^0 - \lambda_1}.$$

We obtain the following expresion for p:

$$\begin{aligned} p &= i\eta^0 + \sqrt{\nu}\Xi\left[\beta_1 - \frac{\delta_{11}}{\eta^0 - \lambda_1} - i\left(\alpha_1 + \frac{\delta_{12}}{\eta^0 - \lambda_1}\right)\right] \\ &= i\left[\eta^0 - \sqrt{\nu}\Xi\left(\alpha_1 + \frac{\delta_{12}}{\eta^0 - \lambda_1}\right)\right] + \sqrt{\nu}\Xi\left(\beta_1 - \frac{\delta_{11}}{\eta^0 - \lambda_1}\right). \end{aligned} \tag{6.30}$$

The stability of the steady rotation is ensured by the condition $\operatorname{Re} p < 0$, i.e.,

$$\operatorname{Re} p = \sqrt{\nu}\Xi\left(\beta_1 - \frac{\delta_{11}}{\eta^0 - \lambda_1}\right) < 0. \tag{6.31}$$

The analysis performed above leads to the following conclusions. First, the presence of viscosity leads to the displacement of the eigenfrequencies (partial frequencies) by a value proportional to $\sqrt{\nu}$, namely, by

$$\sqrt{\nu}\Xi\left(\beta_1 - \frac{\delta_{11}}{\eta^0 - \lambda_1}\right).$$

Secondly, the stability criterion (6.31) in the viscous case differs from that in the case of an ideal fluid, in which the stability criterion is the requirement that the roots of the characteristic equation be real.

Thus, in some cases, viscosity leads to the stabilization of the steady rotation (when A is the greatest moment of inertia), while in other cases, it leads to the loss of stability (when A is the least moment of inertia).

If we consider the Cauchy problem and choose the initial condition to be close to a rotation about the greatest or the least axis of the inertia ellipsoid, then the motion reduces to a uniform rotation and small oscillations about this axis.

The rate of the amplification or attenuation of the joint oscillations of the body and the fluid depends on the ratio of the masses of the fluid and the body, which agrees with the conclusions of [36].

If the fluid density ρ is sufficiently small the free oscillations rapidly decay and the motion is purely forced.

All quantities characterizing the motion of a body with a fluid depend on time as e^{pt}, as in the study of the stability of the steady rotation performed in this book.

7 EQUATIONS OF MOTION FOR A RIGID BODY WITH A CAVITY EQUIPPED WITH FLUID OSCILLATION DAMPERS

This section studies the oscillations of a viscous fluid in a cavity of a rigid body performing a librational motion. A part of the cavity is filled with a viscous fluid, and the other part is filled with a gas at constant pressure. Moreover, the cavity carries constructive elements, such as radial and annular edges. We derive general equations of the perturbed motion of the body with a fluid. Energy dissipation during the oscillation period is set to be small in comparison with the energy of the system.

In the study of the stability of motion of a rigid body with a fluid, an essential role is played by the influence of the viscosity of the components of liquid fuels. There exists two mechanisms of energy dissipation under oscillations of a fluid in a cavity. One of them is related to the vortex formation on the walls of the cavity and to the further energy dissipation in a thin near-wall layer (this occurs in the case of a cavity with smooth walls and large Reynolds numbers), and the other, to the separation of powerful discrete vortices, which afterwards dissipate over the entire volume of the fluid (this occurs in the case of a cavity having constructive elements with acute edges). The latter effect is substantially nonlinear; usually it is at least two orders higher than the boundary layer effect.

This requires including additional dissipative forces in the equations of motion of a rigid body with a fluid-filled cavity.

The method implemented in this chapter is close to a method extensively used by Landau [198]. This is a simple and visual method allowing for the effect of the second mechanism of energy dissipation, which is particularly important for applications.

This section considers a class of perturbed motions of a body with a fluid with small relative energy dissipation, as well as generalized coordinates characterizing the perturbed motion of the rigid body and the fluid. The shape of the cavity is arbitrary. In the case of an ideal fluid, the obtained system of equations transforms into the equations of [148]. The results contain results of [141, 277, 304] as a special case.

Consider a rigid body with a cavity one part of which is filled with a viscous incompressible fluid of density ρ and kinematic viscosity ν and the other part, with a gas at constant pressure. The walls of the cavity carry edges formed by surfaces orthogonal to the surface of the cavity.

We assume that the unperturbed motion is a progressive motion of the body together with the fluid. In the unperturbed motion, the fluid occupies the volume Q bounded by the surface S of the cavity walls and the plane of the free surface Σ perpendicular to the total acceleration vector $\boldsymbol{j}$ of the mass forces in the unperturbed motion (the overload vector).

The perturbed motion of the rigid body is determined by the vectors $\boldsymbol{u}(t)$ of small translational displacement and $\boldsymbol{\Theta}(t)$ of small rotational displacement. The perturbed motion of the fluid is characterized by parameters $S_k(t)$, each of which has the meaning of the amplitude of the kth tone of fluid oscillations at some point of the free surface.

Following [261], we write equations of the perturbed motion of the body with a fluid, neglecting terms of higher orders of smallness in generalized coordinates as

$$\begin{cases} (m^0 + m)\ddot{\boldsymbol{u}} + (L^0 + L)\ddot{\boldsymbol{\Theta}} + \sum\limits_{k=1}^{\infty} \boldsymbol{\lambda}_k \ddot{S}_k = \boldsymbol{P} + \delta \boldsymbol{P}, \\ (J^0 + J)\ddot{\boldsymbol{\Theta}} + (\bar{L}^0 + \bar{L})\ddot{\boldsymbol{u}} - (\boldsymbol{j}, (\tilde{L}^0 + \tilde{L})\boldsymbol{\Theta}) + \sum\limits_{k=1}^{\infty} \boldsymbol{\lambda}_{0k} \ddot{S}_k = \boldsymbol{M}_0 + \delta \boldsymbol{M}_0, \\ \mu_k(\ddot{S}_k + \omega_k^2 S_k) + (\boldsymbol{\lambda}_k, \ddot{\boldsymbol{u}}) + (\boldsymbol{\lambda}_{0k}, \ddot{\boldsymbol{\Theta}}) = \delta P_k \quad (k = 1, 2, \ldots). \end{cases} \tag{7.1}$$

Here m^0 and m are the masses of the rigid body and the fluid, J^0 and J are the symmetric inertia tensors of the body and the associated moments of inertia of the fluid, L^0 and L are the antisymmetric tensors of the static moments of the body and the fluid, $\bar{L}^0$ and $\bar{L}$ are the tensors conjugate to L^0 and L, $\tilde{L}^0$ and $\tilde{L}$ are tensors characterizing the moments of the mass forces acting in the unperturbed motion, $\boldsymbol{\lambda}_k$ and $\boldsymbol{\lambda}_{0k}$ are vectors characterizing the inertial couplings between the motions of the body and the wave motions of the fluid, and $\delta \boldsymbol{P}$, $\delta \boldsymbol{M}_0$, and δP_k are the generalized forces caused by energy dissipation in the cavity.

At $\delta \boldsymbol{P} = \delta \boldsymbol{M}_0 = \delta P_k = 0$, equations (7.1) transform into the equations of the perturbed motion of a body with a cavity partially filled with an ideal fluid obtained in [221]. In what follows, we assume that the coefficients in the equations of motion of a body with an ideal fluid are known.

Thus, the problem of composing equations of motion for a body with a cavity partially filled with a viscous fluid reduces to determining the generalized dissipative forces $\delta \boldsymbol{P}$, $\delta \boldsymbol{M}_0$, and δP_k.

Suppose that a cavity with smooth walls has m edges formed by elements of surfaces orthogonal to the surface S of the cavity. We assume that the width of the edge b counted along the normal to the cavity surface is small in comparison with the characteristic size of the cavity and the least distance between edges. We also assume that the motion of the fluid occurs at large Reynolds numbers ($\mathrm{Re} = Vl/\nu \gg 1$, where V is the characteristic speed of the fluid and l is the characteristic size of the cavity).

In this case, energy dissipates both a thin boundary layer near the boundaries of the region occupied by the fluid (which corresponds to the first dissipation mechanism) and in the entire volume of the fluid (which corresponds to the second dissipation mechanism). The rate of energy dissipation per unit volume of the fluid is of the order of $1/\sqrt{\mathrm{Re}}$ near the wetted surface S of the cavity and of the order of $1/\mathrm{Re}$ near the free surface of the fluid [34].

In what follows, we only retain terms of the order of $1/\sqrt{\mathrm{Re}}$ in expressions for dissipative forces and neglect energy dissipation near the free surface of the fluid.

Since the thickness of the boundary layer is small (of the order of $1/\sqrt{\mathrm{Re}}$), an element of the surface S can be identified with an element of the infinite plane bounding the half-space filled with a viscous fluid and moving at a velocity equal to the difference between the velocities of the rigid body and the perturbed motion of the ideal fluid.

The force acting on an element of the surface S moving at a velocity $V(r,t)$ can be written in the form [296]

$$dF = -\rho\sqrt{\frac{\nu}{\pi}}\int_0^t \frac{\dot{V}(r,\tau)d\tau}{\sqrt{t-\tau}}\,dS, \tag{7.2}$$

where r is the radius vector of the current point of the surface S. Then the rate of energy dissipation in the cavity which occurs by the first dissipation mechanism equals

$$\dot{E} = \int_S (v, dF) = -\rho\sqrt{\frac{\nu}{\pi}}\int_S \left(V(r,t), \int_0^t \frac{\dot{V}(r,\tau)d\tau}{\sqrt{t-\tau}}\,dS\right). \tag{7.3}$$

We represent the velocity v^0 of the perturbed motion of the ideal fluid in the form [148]

$$v^0 = \dot{u} + \sum_{j=1}^{3}\dot{\Theta}_j\nabla\Psi_j + \sum_{k=1}^{\infty}\dot{S}_k\nabla\varphi_k, \tag{7.4}$$

where the $\Theta_j = (\boldsymbol{\Theta}, e_j)$ are the components of the vector $\boldsymbol{\Theta}$ in the body coordinate system $Ox_1x_2x_3$, the e_j are the basis vectors of the system $Ox_1x_2x_3$, the $\Psi_j(x_1,x_2,x_3)$ and the $\varphi_k(x_1,x_2,x_3)$ are solutions of boundary value problems depending only on the geometry of the cavity. Thus, for the relative velocity v, we obtain

$$v = \dot{u} + \dot{\Theta}\times r - V^0 = -\sum_{j=1}^{3}\dot{\Theta}_j(\nabla\Psi_j + r\times e_j) - \sum_{k=1}^{\infty}\dot{S}_k\nabla\varphi_k. \tag{7.5}$$

Moreover, a linear element of an edge can be identified with an element of an infinitely long plate perpendicular to the boundary of the half-space filled with the fluid and oscillating along this boundary. The speed of oscillations equals the normal (to the edge) component of the relative velocity of the ideal fluid in the absence of edges at the points corresponding to the midline of the given edge element.

The component of the relative velocity normal to the edge has the form

$$v_n = (V, n) = -\sum_{j=1}^{3}\dot{\Theta}_j\left(\frac{\partial\Psi_j}{\partial n} - (r\times n)_j\right) - \sum_{k=1}^{\infty}\dot{S}_k\frac{\partial\varphi_k}{\partial n}, \tag{7.6}$$

where $r(\Gamma)$ is the radius vector of a points on the midline of the edge, Γ is the arc length of the midline, and n is the outer normal vector to the surface of the edge, which forms an acute angle with the relative velocity of the fluid. The index j corresponds to the jth component of the vector.

In [220] a semiempirical approach was used, which was based on the representation of the resistance force acting on a linear element of the edge in the form

$$dF = \rho c b^{3/2} \omega V_n \boldsymbol{n}\, d\Gamma, \tag{7.7}$$

where the coefficient c depends on the amplitude of oscillations.

Expression (7.7) is convenient in considering steady harmonic oscillations; however, constructing a dynamic scheme for transition processes on the basis of (7.7) requires additional assumptions.

We suggested a different expression [51], which is free of this drawback:

$$dF = \rho c_x b^{3/2} \int_0^t \frac{\dot{V}_n\, d\tau}{\sqrt{t-\tau}} \boldsymbol{n}\, d\Gamma. \tag{7.8}$$

The coefficient c_x is determined from the condition that the energy dissipation during the oscillation period must be the same in the cases of the resistance laws (7.7) and (7.8).

It is easy to show that expression (7.8) in the mode of harmonic oscillations coincides with (7.7), but since (7.8) does not depend on the amplitude and the frequency of the harmonic process, it can also be used for analyzing transition processes.

The rate of energy dissipation in the cavity in the case of the second dissipation mechanism can be written in the form

$$\dot{E} = \sum_{l=1}^{m} \int_{\Gamma_l} (\boldsymbol{V}, dF), \tag{7.9}$$

where m is the number of edges and Γ_l is the contour formed by the midline of the edge.

Whichever the dissipation mechanism, the rate of energy dissipation in the cavity has the form

$$\dot{E} = -\rho\sqrt{\frac{\nu}{\pi}} \int_S \left(\boldsymbol{V}(\boldsymbol{r},t), \int_0^t \frac{\dot{\boldsymbol{V}}(\boldsymbol{r},\tau)\, d\tau}{\sqrt{t-\tau}} \right) dS + \rho c_x b^{3/2} \sum_{l=1}^{m} \int_{\Gamma_l} \left(\boldsymbol{V}(\boldsymbol{r},t), \int_0^t \frac{\dot{V}_n(\tau) d\tau}{\sqrt{t-\tau}} \boldsymbol{n} \right) d\Gamma. \tag{7.10}$$

The first term in (7.10) describes energy dissipation on the smooth walls, and the second describes dissipation caused by the presence of edges in the cavity. Substituting the expressions (7.5) and (7.6) for the relative velocity of the fluid into (7.10), we obtain

$$\begin{aligned} \dot{E} = &-\left(B \int_0^t \frac{\ddot{\boldsymbol{\Theta}}(\tau) d\tau}{\sqrt{t-\tau}}, \dot{\boldsymbol{\Theta}} \right) - \sum_{k=1}^{\infty} \left(\boldsymbol{\beta}'_{0k} \int_0^t \frac{\ddot{S}_k(\tau) d\tau}{\sqrt{t-\tau}}, \dot{\boldsymbol{\Theta}} \right) \\ &- \sum_{k=1}^{\infty} \left(\boldsymbol{\beta}'_{0k}, \int_0^t \frac{\ddot{\boldsymbol{\Theta}}(\tau) d\tau}{\sqrt{t-\tau}} \right) \dot{S}_k - \sum_{k=1}^{\infty} \left(\sum_{n=1}^{\infty} \beta'_{kn} \int_0^t \frac{\ddot{S}_n(\tau) d\tau}{\sqrt{t-\tau}} \right) \dot{S}_k, \end{aligned} \tag{7.11}$$

where B is the tensor with components $\beta_{ij}^{0'}$, the $\boldsymbol{\beta}'_{0k}$ are vectors with components β'_{0kj}, and the β'_{kn} are scalars.

We have

$$\beta_{ij}^{0'} = \beta_{ji}^{0'} = \rho\sqrt{\frac{\nu}{\pi}}\int\limits_S (\boldsymbol{r}\times\boldsymbol{e}_j + \nabla\Psi_j, \boldsymbol{r}\times\boldsymbol{e}_i + \nabla\Psi_i)dS$$

$$+\rho c_x\sum_{l=1}^{m} b_l^{3/2}\int\limits_{\Gamma_l}\left(\frac{\partial\Psi_j}{\partial\boldsymbol{n}} - (\boldsymbol{r}\times\boldsymbol{n})_j, \frac{\partial\Psi_i}{\partial\boldsymbol{n}} - (\boldsymbol{r}\times\boldsymbol{n})_i\right)d\Gamma, \tag{7.12}$$

$$\beta'_{0kj} = \rho\sqrt{\frac{\nu}{\pi}}\int\limits_S (\boldsymbol{r}\times\boldsymbol{e}_j + \nabla\Psi_j, \nabla\varphi_k)dS + \rho c_x\sum_{l=1}^{m} b_l^{3/2}\int\limits_{\Gamma_l}\left(\frac{\partial\Psi_j}{\partial\boldsymbol{n}} - (\boldsymbol{r}\times\boldsymbol{n})_j, \nabla\varphi_k\right)d\Gamma,$$

$$\beta'_{kn} = \beta'_{nk} = \rho\sqrt{\frac{\nu}{\pi}}\int\limits_S (\nabla\varphi_n, \nabla\varphi_k)dS + \rho c_x\sum_{l=1}^{m} b_l^{3/2}\int\limits_{\Gamma_l}\left(\frac{\partial\varphi_n}{\partial\boldsymbol{n}}, \frac{\partial\varphi_k}{\partial\boldsymbol{n}}\right)d\Gamma,$$

$$(j = 1,2,3,\ \ n,k = 1,2,\ldots,\ \ l = 1,2,\ldots,m).$$

The first term in each formula in (7.12) allows for the boundary layer effect and the second, for the effect of vortex formation on the edges of the edge verges. For finite oscillation amplitudes of the body and the fluid, the second term dominates; the first begins to manifest itself at amplitudes approaching zero (or in the absence of edges).

On the other hand, the rate of energy dissipation is determined by the expression

$$\dot{E} = (\delta\boldsymbol{P}, \dot{\boldsymbol{u}}) + (\delta\boldsymbol{M}_0, \dot{\boldsymbol{\Theta}}) + \sum_{k=1}^{\infty}\delta P_k\dot{S}_k. \tag{7.13}$$

Comparing (7.11) and (7.13), we obtain

$$\delta\boldsymbol{P} = 0,$$

$$\delta\boldsymbol{M}_0 = -B\int\limits_0^t\frac{\ddot{\boldsymbol{\Theta}}(\tau)d\tau}{\sqrt{t-\tau}} - \sum_{k=1}^{\infty}\boldsymbol{\beta}'_{0k}\int\limits_0^t\frac{\ddot{S}_k(\tau)d\tau}{\sqrt{t-\tau}},$$

$$\delta P_k = -\left(\boldsymbol{\beta}'_{0k}, \int\limits_0^t\frac{\ddot{\boldsymbol{\Theta}}(\tau)d\tau}{\sqrt{t-\tau}}\right) - \sum_{n=1}^{\infty}\beta'_{kn}\int\limits_0^t\frac{\ddot{S}_n(\tau)d\tau}{\sqrt{t-\tau}}. \tag{7.14}$$

These relations completely solve the problem of composing equations for the perturbed motion of a body with a cavity partially filled with a viscous fluid at large Reynolds numbers.

Substituting (7.14) into (7.1), we obtain the following equations for the weakly excited motion of a fluid-containing rigid body:

$$
\left\{
\begin{aligned}
&(m^0+m)\ddot{\boldsymbol{u}}+(L^0+L)\ddot{\boldsymbol{\Theta}}+\sum_{k=1}^{\infty}\boldsymbol{\lambda}_k\ddot{S}_k=\boldsymbol{P},\\
&(J^0+J)\ddot{\boldsymbol{\Theta}}+B\int_0^t\frac{\ddot{\boldsymbol{\Theta}}(\tau)d\tau}{\sqrt{t-\tau}}+(\bar{L}^0+\bar{L})\ddot{\boldsymbol{u}}-(\boldsymbol{j},(\tilde{L}^0+\tilde{L})\boldsymbol{\Theta})\\
&+\sum_{k=1}^{\infty}\left(\boldsymbol{\lambda}_{0k}\ddot{S}_k+\boldsymbol{\beta}'_{0k}\int_0^t\frac{\ddot{S}_k(\tau)d\tau}{\sqrt{t-\tau}}\right)=\mathbf{M}_0,\\
&\mu_k(\ddot{S}_k+\omega_k^2S_k)+(\boldsymbol{\lambda}_k,\ddot{\boldsymbol{u}})+(\boldsymbol{\lambda}_{0k},\ddot{\boldsymbol{\Theta}})+\left(\boldsymbol{\beta}'_{0k},\int_0^t\frac{\ddot{\boldsymbol{\Theta}}(\tau)d\tau}{\sqrt{t-\tau}}\right)\\
&=\sum_{n=1}^{\infty}\beta'_{kn}\int_0^t\frac{\ddot{S}_n(\tau)d\tau}{\sqrt{t-\tau}},\quad (k=1,2,\dots).
\end{aligned}
\right.
\tag{7.15}
$$

The system of equations (7.15) can be regarded as Lagrange equations of the second kind in which the u_j, Θ_j, and S_k (where $j=1,2,3$ and $k=1,2,\dots$) play the role of generalized coordinates. System (7.15) makes it possible to consider regimes in which the dynamic system is subject to perturbations being arbitrary functions of time.

8 AN INTEGRAL RELATION IN THE CASE OF A VISCOUS FLUID

We return to the equations (6.1) for the perturbed motion of a fluid-containing rigid body. In the cylindrical coordinate system, these equations have the form

$$
\begin{aligned}
&A\dot{\Omega}+i(C-A)\omega_0\Omega+2\rho\sum_{n=1}^{\infty}a_n(\dot{S}_n-i\omega_0S_n)=M,\\
&\mathrm{N}_n^2\left\{\dot{S}_n-i\lambda_nS_n+\sqrt{\frac{\nu}{\pi}}\int_0^t\frac{\dot{S}_n(\tau)\alpha_n(t-\tau)+S_n(\tau)\beta_n(t-\tau)}{\sqrt{t-\tau}}d\tau\right\}+a_n\dot{\Omega}=0\\
&(n=1,2,\dots),
\end{aligned}
\tag{8.1}
$$

where A and C are the principal moments of inertia about the axes Ox and Oz, respectively, $\Omega=\Omega_x-i\Omega_y$, $M=M_x-iM_y$, and the $\alpha_n(t-\tau)$ and $\beta_n(t-\tau)$ are determined by solving the boundary value problem (6.2) in this chapter. It is taken into account that the cross coefficients α_{mn} and β_{mn} of the inertial couplings weakly affect the dynamics of the rotating body with a viscous fluid, so that we can neglect their influence and retain only the principal terms with $m=n$.

Applying the Laplace transform to all equations in system (8.1), expressing $\hat{S}_n$ from each kth equation ($k \geqslant 2$), and substituting the result into the first equation, we obtain

$$[Ap + i(C - A)\omega_0]\hat{\Omega} - 2\rho \sum_{n=1}^{\infty} \frac{a_n^2(p - i\omega_0)p\hat{\Omega}}{N_n^2 \Psi_n(p)} = \hat{M}(p), \tag{8.2}$$

where

$$\Psi_n(p) = p - i\lambda_n + \frac{A_n}{p}\sqrt{\nu} + \frac{1}{2}\sqrt{\nu}B_n(p + 2\omega_0)\left(\frac{1}{\sqrt{p + i2\omega_0}} + \frac{1}{\sqrt{p - i2\omega_0}}\right)$$
$$- \sqrt{\nu}C_n(p - 2\omega_0)\left(\frac{1}{\sqrt{p + i2\omega_0}} - \frac{1}{\sqrt{p - i2\omega_0}}\right).$$

Characteristic equation of the system in the unperturbed motion ($M = 0$) has the form

$$Ap + i(C - A)\omega_0 - p(p - i\omega_0)\sum_{n=1}^{\infty} \frac{E_n}{\Psi_n(p)} = 0, \quad E_n = \frac{2\rho a_n^2}{N_n^2}. \tag{8.3}$$

We find the roots of the function $\Psi_n(p)$ by the perturbation method. Suppose that $p_n = i\lambda_n + \sqrt{\nu}\delta_n$ and the correction of the first order in $\sqrt{\nu}$ to the root $\delta_n = \delta_n(p_n)$ is determined from the condition $\Psi_n(p_n) = 0$. Taking the Laurent expansion of the function $1/\Psi_n(p)$ and retaining only the terms in neighborhood of poles, we reduce equation (8.3) to the form

$$Ap + i(C - A)\omega_0 - p(p - i\omega_0)\sum_{n=1}^{\infty} \frac{E_n}{\Psi_n'(p_n)(p - p_n)} = 0, \tag{8.4}$$

where $\Psi_n'(p_n) = 1 + \sqrt{\nu}\,\mathrm{Re}\,Q_n + i\sqrt{\nu}\,\mathrm{Im}\,Q_n$ and Q_n is determined by

$$Q_n = Q_n(\lambda_n) = \frac{1-i}{2\sqrt{2}}B_n(q^+ + q^-) + \frac{1-i}{4\sqrt{2}}B_n(q^{+3} - q^{-3}) - \frac{1-i}{\sqrt{2}}C_n(q^+ - q^-)$$
$$+ \frac{1-i}{\sqrt{2}}C_n\omega_0(q^{+3} + q^{-3}) + \frac{A_n}{\lambda_n^2},$$
$$q^+ = \frac{1}{\sqrt{p_n + 2i\omega_0}}, \quad q^- = \frac{1}{\sqrt{p_n - 2i\omega_0}}.$$

The characteristic equation (8.4) can be rewritten in the form

$$Ap + i(C - A)\omega_0 - p(p - i\omega_0) \times \sum_{n=1}^{\infty} \frac{E_n\left(1 - \sqrt{\nu}\,\mathrm{Re}\,Q_n - i\sqrt{\nu}\,\mathrm{Im}\,Q_n\right)}{p - p_n} = 0. \tag{8.5}$$

In the first approximation, instead of the infinite sum, we leave only one principal term ($n = 1$). Making the change of variables $p = i\eta$ and representing p_1 as

$p_1 = i\lambda_1 + \sqrt{\nu}\delta_1$, we obtain

$$A\eta + (C-A)\omega_0 - \frac{\eta(\eta-\omega_0)E_1}{\eta-\lambda_1} \times \left(1 - \sqrt{\nu}\,\mathrm{Re}\,Q_1 - i\sqrt{\nu}\,\mathrm{Im}\,Q_1 + i\frac{\sqrt{\nu}\delta_1}{\eta-\lambda_1}\right) = 0. \tag{8.6}$$

We solve this equation by the perturbation method. Let $\eta_0^{1,2}$ be the roots of the equation (2.7) for the ideal fluid in the stability region. Then we seek the roots of equation (8.6) in the form $\eta^{1,2} = \eta_0^{1,2} + \sqrt{\nu}\Delta^*$. For the viscosity correction Δ^* to the root we obtain

$$\Delta^* = \frac{\eta_0^{1,2}(\eta_0^{1,2}-\omega_0)E_1\left(-\mathrm{Re}\,Q_1 - i\,\mathrm{Im}\,Q_1 + i\frac{\delta_1}{\eta_0^{1,2}-\lambda_1}\right)}{2(A-E_1)\eta_0^{1,2} + (C-A)\omega_0 - A\lambda_1 + \omega_0 E_1},$$

and the roots of (8.5) for $(n=1)$ can be written as

$$p^{1,2} = i\left[\eta_0^{1,2} + \sqrt{\nu}\frac{\eta_0^{1,2}(\eta_0^{1,2}-\omega_0)E_1(-\mathrm{Re}Q_1 - i\,\mathrm{Im}\,Q_1 + i\frac{\delta_1}{\eta_0^{1,2}-\lambda_1})}{2(A-E_1)\eta_0^{1,2} + (C-A)\omega_0 - A\lambda_1 + \omega_0 E_1}\right]. \tag{8.7}$$

Returning to expression (8.2), we obtain the following expression for the angular velocity:

$$\hat{\Omega}(p) = \frac{\hat{M}(p)(p-p_1)}{(A - E_1(1-\sqrt{\nu}Q_1))(p-p^{(1)})(p-p^{(2)})}, \tag{8.8}$$

where $p^{(1)}$ and $p^{(2)}$ are determined by (8.7).

To perform the inverse Laplace transform, we use the convolution formula

$$(f*g)(t) = \int_0^t f(\tau)g(t-\tau)d\tau.$$

In the case under consideration, we have

$$F(p) = \hat{M}(p)$$
$$G(p) = \frac{p-p_1}{(A - E_1(1-\sqrt{\nu}Q_1))(p-p^{(1)})(p-p^{(2)})}.$$

The original of functions of the form $G(p) = A^n(p)/B^m(p)$, where $m > n$, is determined by

$$g(t) = \sum_{k=1}^{m} \frac{A^n(p^{(k)})}{B'^m(p^{(k)})} e^{p^{(k)}t},$$

where the $p^{(k)}$ are the zeros of $B^m(p)$. Returning to (8.8), we obtain

$$g(t) = \sum_{k=1}^{2} \frac{(p^{(k)} - p_1)\, e^{p^{(k)}t}}{2(A - E_1(1 - \sqrt{\nu}Q_1))p^{(k)} + i(\delta\omega_0 - A\lambda_1 + \omega_0 E_1(1 - \sqrt{\nu}Q_1)),}$$

or

$$g(t) = Xe^{p^{(1)}t} + Ye^{p^{(2)}t},$$

where X and Y are determined by

$$X = \frac{p^{(1)} - p_1}{2(A - E_1(1 - \sqrt{\nu}Q_1))p^{(1)} + i(\delta\omega_0 - A\lambda_1 + \omega_0 E_1(1 - \sqrt{\nu}Q_1))},$$

$$Y = \frac{p^{(2)} - p_1}{2(A - E_1(1 - \sqrt{\nu}Q_1))p^{(2)} + i(\delta\omega_0 - A\lambda_1 + \omega_0 E_1(1 - \sqrt{\nu}Q_1))}.$$

The final expression for $\Omega(t)$ as a function of the control moment $M(t)$ in the case of a viscous fluid is

$$\Omega(t) = \int_0^t M(\tau)(Xe^{p^{(1)}(t-\tau)} + Ye^{p^{(2)}(t-\tau)})d\tau. \tag{8.9}$$

Note that equation (8.9) is similar to equation (3.7) of the first chapter for an ideal fluid (with $\nu = 0$), as well as the corresponding expression for $p^{(1,2)}$ and for X and Y.

Expressions (3.7) of the first chapter and (8.9) of this chapter give an explicit dependence of the angular velocity of the perturbed motion of a rigid body with a cavity completely filled with a fluid under the only assumption that the rotation axis of the system in the unperturbed motion is simultaneously an axis of mass and geometric symmetry of the body and the cavity. In the other respects, the shapes of the cavity and the body were arbitrary.

In the next section we treat the moment of external forces as an external control action and use it to change the state of the system, that is, transfer the system from a given state to another desirable state, stabilize the system, control it, and so on. In terms of optimal control theory, we treat the components of the angular velocity of the perturbed motion of the rigid body as phase variables and the external moment as an unknown control function.

9 AN EQUIVALENT SYSTEM IN THE OPTIMAL CONTROL SETTING

Consider expression (8.9). Recall that $\Omega = \Omega_x - i\Omega_y$, $M = M_x - iM_y$, and the values $p^{(1)}$ and $p^{(2)}$ are determined by (8.7). Thus, for each of the components Ω_x and Ω_y, we can write an expression with real coefficients:

$$\Omega_x(t) = \int_0^t M_x(\tau)e^{\mathrm{Re}\,p^{(1)}(t-\tau)}[\mathrm{Re}X\cos\mathrm{Im}\,p^{(1)}(t-\tau) - \mathrm{Im}\,X\sin\mathrm{Im}\,p^{(1)}(t-\tau)]d\tau$$
$$+ \int_0^t M_x(\tau)e^{\mathrm{Re}\,p^{(2)}(t-\tau)}[\mathrm{Re}Y\cos\mathrm{Im}\,p^{(2)}(t-\tau) - \mathrm{Im}\,Y\sin\mathrm{Im}\,p^{(2)}(t-\tau)]d\tau$$
$$+ \int_0^t M_y(\tau)e^{\mathrm{Re}\,p^{(1)}(t-\tau)}[\mathrm{Re}X\sin\mathrm{Im}\,p^{(1)}(t-\tau) + \mathrm{Im}\,X\cos\mathrm{Im}\,p^{(1)}(t-\tau)]d\tau$$
$$+ \int_0^t M_y(\tau)e^{\mathrm{Re}\,p^{(2)}(t-\tau)}[\mathrm{Re}Y\sin\mathrm{Im}\,p^{(2)}(t-\tau) + \mathrm{Im}\,Y\cos\mathrm{Im}\,p^{(2)}(t-\tau)]d\tau, \quad (9.1)$$

$$\Omega_y(t) = -\int_0^t M_x(\tau)e^{\mathrm{Re}\,p^{(1)}(t-\tau)}[\mathrm{Re}X\sin\mathrm{Im}\,p^{(1)}(t-\tau) + \mathrm{Im}\,X\cos\mathrm{Im}\,p^{(1)}(t-\tau)]d\tau$$
$$- \int_0^t M_x(\tau)e^{\mathrm{Re}\,p^{(2)}(t-\tau)}[\mathrm{Re}Y\sin\mathrm{Im}\,p^{(2)}(t-\tau) + \mathrm{Im}\,Y\cos\mathrm{Im}\,p^{(2)}(t-\tau)]d\tau$$
$$+ \int_0^t M_y(\tau)e^{\mathrm{Re}\,p^{(1)}(t-\tau)}[\mathrm{Re}X\cos\mathrm{Im}\,p^{(1)}(t-\tau) - \mathrm{Im}\,X\sin\mathrm{Im}\,p^{(1)}(t-\tau)]d\tau$$
$$+ \int_0^t M_y(\tau)e^{\mathrm{Re}\,p^{(2)}(t-\tau)}[\mathrm{Re}Y\cos\mathrm{Im}\,p^{(2)}(t-\tau) - \mathrm{Im}\,Y\sin\mathrm{Im}\,p^{(2)}(t-\tau)]d\tau. \quad (9.2)$$

Let us introduce the notation

$$A(t) = \int_0^t M_x(\tau)e^{\mathrm{Re}\,p^{(1)}(t-\tau)}[\mathrm{Re}X\cos\mathrm{Im}\,p^{(1)}(t-\tau) - \mathrm{Im}\,X\sin\mathrm{Im}\,p^{(1)}(t-\tau)]d\tau,$$

$$B(t) = \int_0^t M_x(\tau)e^{\mathrm{Re}\,p^{(2)}(t-\tau)}[\mathrm{Re}Y\cos\mathrm{Im}\,p^{(2)}(t-\tau) - \mathrm{Im}\,Y\sin\mathrm{Im}\,p^{(2)}(t-\tau)]d\tau,$$

$$C(t) = \int_0^t M_y(\tau)e^{\mathrm{Re}\,p^{(1)}(t-\tau)}[\mathrm{Re}X\sin\mathrm{Im}\,p^{(1)}(t-\tau) + \mathrm{Im}\,X\cos\mathrm{Im}\,p^{(1)}(t-\tau)]d\tau,$$

$$D(t) = \int_0^t M_y(\tau)e^{\mathrm{Re}\,p^{(2)}(t-\tau)}[\mathrm{Re}Y\sin\mathrm{Im}\,p^{(2)}(t-\tau) + \mathrm{Im}\,Y\cos\mathrm{Im}\,p^{(2)}(t-\tau)]d\tau.$$

Similar expressions can be obtained for $E(t)$, $F(t)$, $G(t)$, and $H(t)$. Since $\Omega_x(t) = A(t)+B(t)+C(t)+D(t)$ and $\Omega_y(t)=E(t)+F(t)+G(t)+H(t)$, we arrive at the following system equivalent to (8.9):

$$\dot{x}(t) = \begin{cases} \dot{\Omega}_x(t) = \dot{A}(t)+\dot{B}(t)+\dot{C}(t)+\dot{D}(t), \\ \dot{\Omega}_y(t) = \dot{E}(t)+\dot{F}(t)+\dot{G}(t)+\dot{H}(t), \\ \dot{A}(t) = M_x(t)\,\mathrm{Re}\,X + \mathrm{Re}\,p^{(1)}A(t) - \mathrm{Im}\,p^{(1)}E(t), \\ \dot{B}(t) = M_x(t)\,\mathrm{Re}\,Y + \mathrm{Re}\,p^{(2)}B(t) - \mathrm{Im}\,p^{(2)}F(t), \\ \dot{C}(t) = M_y(t)\,\mathrm{Im}\,X + \mathrm{Re}\,p^{(1)}C(t) + \mathrm{Im}\,p^{(1)}G(t), \\ \dot{D}(t) = M_y(t)\,\mathrm{Im}\,Y + \mathrm{Re}\,p^{(2)}D(t) + \mathrm{Im}\,p^{(2)}H(t), \\ \dot{E}(t) = M_x(t)\,\mathrm{Im}\,X + \mathrm{Re}\,p^{(1)}E(t) + \mathrm{Im}\,p^{(1)}A(t), \\ \dot{F}(t) = M_x(t)\,\mathrm{Im}\,Y + \mathrm{Re}\,p^{(2)}F(t) + \mathrm{Im}\,p^{(2)}B(t), \\ \dot{G}(t) = M_y(t)\,\mathrm{Re}\,X + \mathrm{Re}\,p^{(1)}G(t) - \mathrm{Im}\,p^{(1)}C(t), \\ \dot{H}(t) = M_y(t)\,\mathrm{Re}\,Y + \mathrm{Re}\,p^{(2)}H(t) - \mathrm{Im}\,p^{(2)}D(t); \end{cases} \qquad x(0) = x_0. \tag{9.3}$$

Let us introduce the following notation ($n = 10$, $m = 2$): x_0 is the zero column of height n,

$$A_{n\times n} = \begin{pmatrix} 0 & 0 & \mathrm{Re}\,p^{(1)} & \mathrm{Re}\,p^{(2)} & \mathrm{Re}\,p^{(1)} & \mathrm{Re}\,p^{(2)} & -\mathrm{Im}\,p^{(1)} & -\mathrm{Im}\,p^{(2)} & \mathrm{Im}\,p^{(1)} & \mathrm{Im}\,p^{(2)} \\ 0 & 0 & \mathrm{Im}\,p^{(1)} & \mathrm{Im}\,p^{(2)} & -\mathrm{Im}\,p^{(1)} & -\mathrm{Im}\,p^{(2)} & \mathrm{Re}\,p^{(1)} & \mathrm{Re}\,p^{(2)} & \mathrm{Re}\,p^{(1)} & \mathrm{Re}\,p^{(2)} \\ 0 & 0 & \mathrm{Re}\,p^{(1)} & 0 & 0 & 0 & -\mathrm{Im}\,p^{(1)} & 0 & 0 & 0 \\ 0 & 0 & 0 & \mathrm{Re}\,p^{(2)} & 0 & 0 & 0 & -\mathrm{Im}\,p^{(2)} & 0 & 0 \\ 0 & 0 & 0 & 0 & \mathrm{Re}\,p^{(1)} & 0 & 0 & 0 & \mathrm{Im}\,p^{(1)} & 0 \\ 0 & 0 & 0 & 0 & 0 & \mathrm{Re}\,p^{(2)} & 0 & 0 & 0 & \mathrm{Im}\,p^{(2)} \\ 0 & 0 & \mathrm{Im}\,p^{(1)} & 0 & 0 & 0 & \mathrm{Re}\,p^{(1)} & 0 & 0 & 0 \\ 0 & 0 & 0 & \mathrm{Im}\,p^{(2)} & 0 & 0 & 0 & \mathrm{Re}\,p^{(2)} & 0 & 0 \\ 0 & 0 & 0 & 0 & -\mathrm{Im}\,p^{(1)} & 0 & 0 & 0 & \mathrm{Re}\,p^{(1)} & 0 \\ 0 & 0 & 0 & 0 & 0 & -\mathrm{Im}\,p^{(2)} & 0 & 0 & 0 & \mathrm{Re}\,p^{(2)} \end{pmatrix},$$

$$B_{n\times m} = \begin{pmatrix} \mathrm{Re}(X+Y) & \mathrm{Im}(X+Y) & \mathrm{Re}\,X & \mathrm{Re}\,Y & 0 & 0 & \mathrm{Im}\,X & \mathrm{Im}\,Y & 0 & 0 \\ \mathrm{Im}(X+Y) & \mathrm{Re}(X+Y) & 0 & 0 & \mathrm{Im}\,X & \mathrm{Im}\,Y & 0 & 0 & \mathrm{Re}\,X & \mathrm{Re}\,Y \end{pmatrix}^T. \tag{9.4}$$

Taking into account (9.4), we rewrite system (9.3) in the form

$$\begin{cases} \dot{x}(t) = Ax(t) + BM(t), \\ x(0) = x_0. \end{cases} \tag{9.5}$$

Note that system (9.5) has the same form as system (4.5) in the first chapter with the only difference being that its order for a viscous fluid is higher. As a result, we obtain a system of ten equations.

To conclude this chapter, we make several remarks. The mathematical models of the body–fluid system are equations of the perturbed motion of these systems in generalized coordinates of two types, which characterize the wave motion of the fluid. These equations can be obtained either as Lagrange equations of the second kind or by using theorems on the change of the linear and kinetic momenta of the body–fluid system and the constancy of pressure on the free surface of the fluid.

Let us dwell on the second method in more detail. To implement this method, we must possess the field of absolute velocities of the perturbed motion of the fluid and a relation between the displacement field of its particles and the field of accelerations on the free surface at constant pressure. Applying the theorem about the change of the linear and kinetic momenta with respect to the point O for the body–fluid system, we obtain a general system of equations for the perturbed motion of the body with the fluid. Moreover, the right-hand side of the equations for the kinetic momentum contains the moment of the system of mass forces in the unperturbed motion, which is determined by the rotation of the body and by the change of the fluid free surface configuration under the perturbed motion of the body. As a result, we obtain two mathematical models for the body–fluid system, a scheme with a "floating lid" and a scheme with a "fixed lid." These mathematical models are different in form but completely equivalent in essence. It can be shown that one of them is transformed into the other by a nonsingular linear transformation.

We emphasize that the mathematical models for vortex phenomena presented in the book adequately describe the dynamical processes provided that the turbulent motions of the system are already formed; thus, they do not generally apply to describe the processes of vortex emergence and disappearance.

Conclusion

This book is devoted to the study of the dynamics of rotating bodies with fluid-containing cavities for the two main classes of motion of fluid-containing bodies, rotational and librational, which is in the channel of the most important applications. The case of the complete filling of the cavity with an ideal or a viscous fluid and the case of partial filling for the rotational motion of the rigid body were considered.

In the book, methods for obtaining relations between angular velocities perpendicular to the main axis of rotation and the moment of external forces treated as a control were presented. An attempt to bulid a bridge from optimal control theory to the dynamics of fluid-containing rotating bodies was made. This first step demonstrates the possibility of setting and solving control problems for a class of systems interesting and important for mechanics.

One could say that everything is in place to tackle, in particular, the following problem? Suppose that a fluid-containing rotating rigid body has a longitudinal angular velocity at some moment of time. It is required to nullify this velocity by means of control moments as fast as possible, thereby suppressing the undesirable longitudinal motions of the rotor.

The authors hope that this study will serve as a starting point and stimulus for considering other problems in the dynamics of fluid-filled rotating rigid bodies and, moreover, make it possible to state and analyze particular optimal control problems in this area.

References

[1] Abramson H. N. Dynamic Behavior of Liquid in Moving Containers. Appl. Mech. Review, vol. 16, no. 7, 1963.

[2] Abramson H. N., Chu W. H., and Ransleben G. E. Representation of Fuel Sloshing in Cylindrical Tanks by Equivalent Mechanical Model. ARS Journal, 1961, vol. 16, no. 12.

[3] Abramson H. N., Gazza L. R., and Kana D. D. Liquid Sloshing in Compartmented Cylindrical Tanks. ARS Journal, 1962, vol. 32, no. 6.

[4] Aizerman M. A. 1980. Classical Mechanics. Moscow: Fizmatgiz, 1980 [in Russian].

[5] Akulenko L. D. Asymptotic Methods of Optimal Control. Moscow: Nauka, 1987 [in Russian].

[6] Akulenko L. D. The perturbed time-optimal problem of controlling the final position of a material point by means of a limited force Point by Means of a Bounded Force // Prikl. Mat. Mekh., 1994, vol. 58. no. 2. pp. 12–21 [J. Appl. Math. Mech. 1994, vol. 58, no. 2, 197–206].

[7] Akulenko L. D. Approximate synthesis optimal control motion over part variables // Izv. Akad. Nauk. Mekh. Tverd. Tela, 1980, no. 5, pp. 3–13 [in Russian].

[8] Akulenko L. D. Problems and Methods of Optimal Control. Dordrecht: Kluwer Acad. Publ., 1994.

[9] Aleksandrov V. V. and Shmyglevskii Yu. D. On inertial and shear flows // Dokl. Akad. Nauk SSSR, 1988, vol. 274, no. 2. pp. 280–283 [in Russian].

[10] Aleksandryan R. A. Spectral Properties of Operators Generated by Systems of Differential Equations of the Type of S. L. Sobolev. Tr. Mosk. Mat. O-va, vol. 12, no. 1, 1960 [in Russian].

[11] Aleksin V. A. and Kazeikin S. N. Modeling the effect of freestream turbulence parameters on unsteady boundary layer flow // Izv. Ross. Akad. Nauk. Mekh. Zhidk. Gaz., 2000, no. 6, pp. 64–77 [Fluid Dyn., 2000, vol. 35, no. 6, pp. 846–857].

[12] Afanas'ev K. E. and Stukolov S. V. Circulation steady plane-parallel flow over a heavy liquid of finite depth with a free surface // Prikl. Mekh. Tech. Fiz., 2000, vol. 41, no. 3, pp. 101–110 [J. Appl. Mech. Tech. Phys., 2000, vol. 41, no. 3, pp. 470–478].

[13] Armstrong G. L. and Kachigan H. Stability and Control of Carrier Vehicles, Handbook of Asronautical Engineering. McGraw Hill Book, New York, 1961.

[14] Bayer H. F. Stability Boundaries of Liquid Propelled Space Vehicles with Sloshing. AIAA Journal, 1963, vol. 1, no. 7.

[15] Bagaeva N. Ya. and Moiseev N. N. Three problems on the vibrations of a viscous fluid, Zh. Vych. Mat. Mat. Fiz., 1964, vol. 4, no. 2, pp. 317-326 [USSR Comput. Math. Math. Phys., 1964, vol. 4, no. 2, pp. 144–158].

[16] Bakanov S. P. and Roldugin V. N. To the question about gas diffusion sliding gas // Inzh.-Fiz. Zh., 1981, vol. 40, no. 5, pp. 807–817 [in Russian].

[17] Batishchev V. A. and Khoroshunova E. V. The formation of rotational regimes in the thermocapillary flow of a non-uniform fluid in a layer // Izv. Ross. Akad. Nauk. Prikl. Mat. Mekh., 2000, vol. 64, no. 4. pp. 560–573 [J. Appl. Math. Mech., 2000, vol. 64, no. 4, pp. 537–545].
[18] Bellman R. Dynamic Programming. Princeton, N.J.: Princeton Univ. Press, 1957.
[19] Bellman R. Adaptive Control Processes, N.J.: Princeton Univ. Press, 1961.
[20] Bellman R. E., Glicksberg I. L., and Gross O. A. Some Aspects of the Mathematical Theory of Control Processes, RAND Corporation, R-313, 1958.
[21] Bellman R. and Dreyfus S. Applied Dynamic Programming, RAND Corporation, R-352-PR, 1962.
[22] Belonosov S. M. and Chernous K. A. Boundary Value Problems for Navier–Stokes Equations. Moscow: Nauka, 1981 [in Russian].
[23] Belyaeva M. A., Myshkis A. D., and Tyuntsov A. D. Hydrostatics in weak gravitational fields: Equilibrium shapes of fluid surface, Izv. Akad. Nauk SSSR, Mekh. Mashinostr., no. 5, 1964 [in Russian].
[24] Bogolyubov N. N. and Mitropol'skii Yu. A. Asymptotic Methods in the Theory of Nonlinear Oscillations, Moscow: Gostekhteorizdat, 1955 [in Russian].
[25] Bogoryad I. B. To the solution of the problem on the oscillations of a fluid partially filling a cavity by the variational method. Prikl. Mat. Mekh., 1962, vol. 26, no. 6 [in Russian].
[26] Boltyanskii V. G., Gamkrelidze R. V., and Pontryagin L. S. To the theory of optimal processes // Dokl. Akad. Nauk SSSR, 1956, vol. 110, no. 1, pp. 7–10 [in Russian].
[27] Bondarenko Yu. A. Inertial three-dimensional motion of nonviscous incompressible fluid // Vopr. Atom. Nauki Tekhn. Ser. Mat. Model. Fiz. Protsessov. 1994, no. 3, pp. 41–46 [in Russian].
[28] Boyarshina L. G. and Koval'chuk P. S. Analysis of nonlinear wave motions of a fluid in a cylindrical vessel experiencing a given angular motion // Prikl. Mekh. (Kiev), 1990, no. 6. pp. 95–101 [in Russian].
[29] Bretherton F. P., Carrier G. F., and Longuet-Higgins M. S. Report of the I.U.T.A.M. symposium on rotating fluid systems. Journal of fluid mech., vol. 26, part 2, 1966.
[30] Bykhovskii E. B. and Smirnov N. V. Orthogonal decomposition of the space of vector functions square-summable on a given domain and operators of vector analysis // Tr. Mat. Inst. im. V. A. Steklova Akad. Nauk SSSR, 1960, vol. 59, pp. 5–36 [in Russian].
[31] Case K. M. and Parkinson W. C. Damping of Surface Waves in an Incompressible Liquid. J. of Fluid Mech., 1957, March, vol. 2, p. 2.
[32] Castro R. Asymptotic form of the solution of the problem of waves on the surface of a heterogeneous stratified liquid // Zh. Vychisl. Mat. Mat. Fiz., 1982, vol. 22, no. 4. pp. 891–902 [U.S.S.R. Comput. Math. Math. Phys. 1982, vol. 22, no. 4, pp. 125–137].
[33] Chernous'ko F. L. Self-similar motion of a fluid under the action of surface tension // Prikl. Mat. Mekh., vol. 29, no. 1, 1965 [in Russian].
[34] Chernous'ko F. L. Rotational motions of a solid body with a cavity filled with fluid // Izv. Akad. Nauk SSSR, Prikl. Mat. Mekh., vol. 31, no. 3, pp. 416–432, 1967 [in Russian].
[35] Chernous'ko F. L. Motion of a solid body with cavities filled with a viscous fluid at small Reynolds numbers // Zh. Vych. Mat. Mat. Fiz., 1965, vol. 5. no. 6. pp. 1049–1070 [in Russian].
[36] Chernous'ko F. L. Motion of a Solid Body with Cavities Containing Viscous Fluid. Moscow: Vychisl. Tsentr, Akad. Nauk SSSR, 1968 [in Russian].
[37] Chernous'ko F. L. Motion of a body with a cavity filled with a viscous fluid at large Reynolds number // Prikl. Mat. Mekh., 1966, vol. 30. no. 3. pp. 476–494 [PMM, J. Appl. Math. Mech. 1966, vol. 30, pp. 568–589].
[38] Chernous'ko F. L. Motion of a fluid thin layer under that action of gravity and surface tension forces // Prikl. Mat. Mekh., vol. 29, no. 5, 1965 [in Russian].

[39] Chernous'ko F. L. Oscillations of a vessel with a viscous fluid // Izv. Akad. Nauk SSSR, Mekh. Zhidk. Gaza, no. 1, 1967 [in Russian].
[40] Chernous'ko F. L. Oscillations of a solid body with a cavity filled with a viscous fluid. // Mekh. Tverd. Tela. 1967, no. 1, pp. 3–14 [in Russian].
[41] Chernous'ko F. L. On the motion of a solid body with a cavity containing an ideal fluid and an air bubble // Prikl. Mat. Mekh., vol. 28, no. 4, 1964 [in Russian].
[42] Chernous'ko F. L. Motion of a body with a cavity partially filled with a viscous fluid // Prikl. Mat. Mekh., 1966, vol. 30, no. 6, pp. 476–494 [in Russian].
[43] Chernous'ko F. L. On the free oscillations of a viscous fluid in a vessel // Prikl. Mat. Mekh., 1966. vol. 30, no. 5, pp. 836–847 [in Russian].
[44] Chernous'ko F. L. Estimation of the Phase State of Dynamical Systems. The Method of Ellipsoids // Moscow: Nauka, 1988 [in Russian].
[45] Chetaev N. G. On the stability of rotational motions of a solid body with a cavity filled with an ideal fluid // Prikl. Mat. Mekh., vol. 21, no. 2, 1957.
[46] Chetaev N. G. Stability of Motion // Moscow: Gostekhizdat, 1946 [in Russian].
[47] Chu W. H. Sloshing of Liquid in Cylindrical Tanks of Elliptical Cross-Section. ARS Journal, 1960, vol. 30, no. 4.
[48] Collatz L. Eigenvalue problems // Handbook of Engineering Mechanics, Chap. 18, New York: McGraw-Hill, 1962.
[49] Cooper R. M. Dynamics of liquids in moving containers. ARS Journal, vol. 30, no. 8, 1960.
[50] Delgado A., Petri B., and Rath H. Fluid management in space by rotating disks // Appl. Microgravity Technol. 1988. no. 4. pp. 188–201.
[51] Desyatov V. T. Experimental Study of the Stability of the Rotational Motion of Fluid-Filled Bodies // Dynamics of Spacecrafts and Investigation of cosmic space. Moscow: Mashinostroenie, 1986, pp. 254–260 [in Russian].
[52] Ditkin V. A. and Prudnikov A. P. Integral Transformations and Operational Calculus, Moscow: Nauka, 1974 [in Russian].
[53] Doetsch G. Guide to the Applications of the Laplace and Z-Transforms, London: Van Nostrand-Reinhold, 1971.
[54] Doetsch G. Anleitung zum Praktischen Gebrauch der Laplace-Transformation und der Z-Transformation. Munchen; Wien: R. Oldenburg, 1967.
[55] Dokuchaev L. V. Nonlinear dynamics and analysis of instability regions of the rotation of deformable artificial Earth satellites by the root method // Izv. Ross. Akad. Nauk. Mekh. Tverd. Tela, 2000. no. 4. pp. 3–17 [in Russian].
[56] Dokuchaev L. V. Nonlinear Dynamics of Aircrafts with Deformable Elements. Moscow: Mashinostroenie, 1987 [in Russian].
[57] Dokuchaev L. V. and Pvalov R. V. On the stability of the steady rotation of a solid body with a fluid-containing cavity // Izv. Akad. Nauk SSSR, Mekh. Tverd. Tela. 1973, no. 2, pp. 6–15 [in Russian].
[58] Dokuchaev L. V., Pvalov R. V., and Rogovoi V. M. Dynamics of a Rotating Body with a Fluid Experiencing Turbulent Motion // Izv. Akad. Nauk SSSR, Mekh. Tverd. Tela, no. 1, 1972 [in Russian].
[59] Dudin G. N. Gas injection from the surface of a triangular plate in a hypersonic flow // Izv. Ross. Akad. Nauk. Mekh. Zhidk. Gaz., 2000. no. 1. pp. 125–138 [Fluid Dyn. 2000, vol. 35, no. 1, pp. 101–107].
[60] Dynnikova G. Ya. An analog of Bernoulli and Cauchy-Lagrange integrals for a time-dependent vortex flow of an ideal incompressible fluid // Izv. Ross. Akad. Nauk. Mekh. Zhidk. Gaz., 2000. no. 1. pp. 31–44 [Fluid Dyn. 2000, vol. 35, no. 1, pp. 24–32].
[61] D'yachenko V. P. Oscillations of a gyroscope partially filled with a fluid // Dokl. Akad. Nauk Ukr. SSR, Ser. A, no. 10, 1971 [in Russian].

[62] Efros A. M. and Danilevskii A. M. Operation Calculus and Contour Integrals. Kharkov: GNTIU, 1937 [in Russian].

[63] Escudier M. Vortex breakdown: observations and explanations // Progr. Aerospace Sci. 1988. vol. 25. no. 2. pp. 189–229.

[64] Faller N. J. and Kaylor R. Unsteady multiple boundary layers on a porous plate in a rotating system. Journal Phys. of Fluid, vol. 16, no. 9, 1973.

[65] Fedoryuk M. V. The Saddle-Point Method. Moscow: Nauka, 1977 [in Russian].

[66] Feynman R. P., Leighton R. B., and Sands M. The Feynman Lectures on Physics, vol. 2, Reading, Mass.: Addison-Wesley, 1964.

[67] Furasov V. D. Stability of Motion, Estimation, and Stabilization. Moscow: Nauka, 1977 [in Russian].

[68] Gabov S. A. Spectrum and bases of eigenfunctions for one problem on acoustic oscillations of a rotating fluid // Dokl. Akad. Nauk SSSR, 1980, vol. 254, no. 4. pp. 777–779 [in Russian].

[69] Gabov S. A., Malysheva G. Yu., Sveshnikov A. G., and Shagov A. K. On some equations arising in the dynamics of a rotating stratified and compressible fluid // Zh. Vych. Mat. Mat. Fiz., 1984, vol. 24, no. 12. pp. 1850–1863 [USSR Comput. Math. Math. Phys., 1984, vol. 24, no. 6, pp. 162–170].

[70] Gabov S. A., Ruban P. I., and Sekerzh-Zen'kovich S. Ya. Diffraction of Kelvin waves by a half-infinite wall in a semibounded basin // ZhVM and MF, 1975, vol. 15, no. 6. pp. 1521–1524 [USSR Comput. Math. Math. Phys., 1975, vol. 15, no. 6, pp. 144–157].

[71] Gabov S. A. and Sveshnikov A. G. Problems of Stratified Fluid Dynamics. Moscow: Nauka, 1986 [in Russian].

[72] Ganiev O. R. and Khabeev N. S. Dynamics and heat and mass transfer of a bubble containing an evaporating drop // Izv. Ross. Akad. Nauk. Mekh. Zhidk. Gaza, 2000. no. 5. pp. 88–97 [Fluid Dyn. 2000, vol. 35, no. 5, pp. 702–708].

[73] Gantmacher F. R. The Theory of Matrices, New York: Chelsea, 1959.

[74] Gataulin I. G. and Stolbetsov V. N. Estimates of coefficients in equations for a perturbed motion of a fluid-containing body // Mekh. Tverd. Tela, 1966, no. 3 [in Russian].

[75] Gelig A. Kh., Leonov G. A., and Yakubovich V. A. Stability of Nonlinear Systems with a Nonunique Equilibrium State. Moscow: Nauka, 1978 [in Russian].

[76] Gourchenkov A. A. A Model of the viscous fluid unsteady flow in a rotating slot. Journal SAMS, 2001, vol. 40, pp. 447–458.

[77] Gourchenkov A. A. The stability of rotating coupled solid and fluid // Intern. Congress on numerical methods in engineering and applied sciences. Universidad de Consepcion, 1992. Proceeding, pp. 253–261, Chile.

[78] Gourchenkov A. A. and Tsurkov V. I. Hydromechanical interaction of viscous fluid and rotating body. International conference on optimization: techniques and application. UK University of Swansea, 1993, July 11–14, U.K.

[79] Gourchenkov A. A. and Tsurkov V. I. The stability problem of rotating coupled solid and fluid. International conference on optimization: techniques and application. UK University of Swansea, 1993, July 11–14, U.K.

[80] Gradshtein I. S. and Ryzhik I. M. Tables of Integrals, Sum, Series, and Products. Moscow: Fizmatgiz, 1962 [in Russian].

[81] Greenhill A. G. On the general motion of a liquid ellipsoid. Proc Camb. Phil. Soc, vol. 4, 1880.

[82] Greensite A. L. Analysis of Liquid–Propellent Mode Stability of a Multitank Ballistic Booster Vehicle. J. of the Aero/Space Sciences. February, 1962.

[83] Greenspan H. P. The Theory of Rotating Fluids, Cambridge: Cambridge University Press, 1968.

[84] Greenspan H. P. and Howard L. N. On a time dependent motion of a rotating fluid. Journal of fluid mech., vol. 17, part 3, 1963.

[85] Greenspan H. P. On almost rigid rotations. Part 2. Journal of fluid mech., vol. 26, part 1, 1966.
[86] Greenspan H. P. On the general theory of contained rotating fluid motions. Journal of fluid mech., vol. 22, part 3, 1965.
[87] Greenspan H. P. On the transient motion of a contained rotating fluid. Journal of fluid mech., vol. 20, part 4, 1964.
[88] Gubkina E. V. and Monakhov V. N. Filtration of a fluid with free boundaries in unbounded redions // Prikl. Mekh. Techn. Fiz., 2000, vol. 41, no. 5, pp. 188–195 [in Russian].
[89] Gupta A. S. Ekman layers on a porous plate. Journal Phys. Fluids, vol. 5, no. 5, 1972.
[90] Gurchenkov A. A. Turbulent motion of a viscous fluid in a cavity of a rotating body, Doctoral Diss. in physics and mathematics, Moscow: 2001 [in Russian].
[91] Gurchenkov A. A. Turbulent Motion of a Fluid in a Cavity of a Rotating Body. Moscow: Mashinostroenie, 2001 [in Russian].
[92] Gurchenkov A. A. Motion of a Body in a Viscous Fluid at large Reynolds Numbers. Moscow: Mosk. Gos. Oblast. Univ., 2005 [in Russian].
[93] Gurchenkov A. A. Whirling Fluid Dynamics in a Cavity of a Rotating Body. Moscow: Fizmatlit, 2010 [in Russian].
[94] Gurchenkov A. A. Dynamics of the Rotational Motion of a Body with a Fluid-Containing Cavity. Candidate's diss. in physics and mathematics, Moscow: 1982 [in Russian].
[95] Gurchenkov A. A. Inertial oscillations of a viscous fluid in a rotating vessel // Physics of Aerodisperse Systems and Physical Kinetics, Moscow: Mosk. Oblast. Pedagog. Univ im. N. K. Krupskoi, 1979, pp. 207–213 [in Russian].
[96] Gurchenkov A. A. Model problems in optimal control theory for dynamical systems with distributed parameters // XVII All-Russia Workshop "Modern Problems of Mathematical Modeling," Dyurso: 2007, p. 192 [in Russian].
[97] Gurchenkov A. A. Internal friction moment of a rapidly rotating cylindrical vessel filled with a viscous fluid // Izv. Vyssh. Uchebn. Zaved. Priborostroenie, 2001 [in Russian].
[98] Gurchenkov A. A. A nonstationary flow of a viscous fluid on a rotating plate // XVI All-Russia conference "Theoretical Foundations and Construction of Numerical Algorithms and Solution of Problems of Mathematical Physics with Application to Multiprocessor Systems," Dyurso, 2006, p. 92 [in Russian].
[99] Gurchenkov A. A. Unsteady flow on a porous plate in the presence of medium injection // Prikl. Mekh. Techn. Fiz., no. 4, 1980 [in Russian].
[100] Gurchenkov A. A. Unsteady motion of a viscous fluid between rotating parallel walls // Prikl. Mekh. Techn. Fiz., 2001, no. 4. pp. 48–51 [in Russian].
[101] Gurchenkov A. A. Unsteady motion of a viscous liquid between rotating parallel walls in the presence of a crossflow // Prikl. Mekh. Tech. Fiz., 2001, no. 4. pp. 48–51 [J. Appl. Mech. Tech. Phys. 2001, vol. 42, no. 4, pp. 603–606].
[102] Gurchenkov A. A. The unsteady motion of a viscous fluid between rotating parallel walls // Prikl. Mat. Mekh., 2002, vol. 66, no. 2, pp. 251–255 [J. Appl. Math. Mech. 2002, vol. 66, no. 2, pp. 239–243].
[103] Gurchenkov A. A. Unsteady boundary layers on the porous plates in a rotating slot with injection and suction // Zh. Vychisl. Mat. Mat. Fiz. 2001, vol. 41, no. 3. pp. 443–449 [Comput. Math. Math. Phys. 2001, vol. 41, no. 3, pp. 413–419].
[104] Gurchenkov A. A. Data processing and hypothesis testing in dynamic systems // Izv. Ross. Akad. Nauk. Teor. Sist. Upr., no. 2, 2000 [in Russian].
[105] Gurchenkov A. A. Velocity field of a viscous fluid in a rotating cylinder // Selected Problems of Physical Kinetics and Hydrodynamics of Dispersion Systems, Available from VINITI, no. 2675-V87, Moscow, 1987 [in Russian].
[106] Gurchenkov A. A. Equations of rotational motion of a solid body with a cavity containing a viscous fluid and inertial coupling coefficients // Physical Kinetics and Hydromechanics

of Dispersion Systems, Available from VINITI, no. 5321-V86, Moscow, 1986 [in Russian].
[107] Gurchenkov A. A. Stability of a fluid-filled gyroscope // Inzh.-Fiz. Zh. 2002, vol. 75, no. 3 [in Russian].
[108] Gurchenkov A. A. and Kostikov A. A. Control of the motion of a fluid-filled top // XVII All-Russia conference "Theoretical Foundations and Construction of Numerical Algorithms and Solution of Problems of Mathematical Physics with Application to Multiprocessor Systems" dedicated to the memory of K. I. Babenko, Abrau–Dyurso, September 15–21 2008, p. 94 [in Russian].
[109] Gurchenkov A. A., Bashlykov A. M., and Gridina E. D. Stability criteria for dynamic systems with distributed parameters // Proceedings of XX All-Russia Workshop "Analytical Methods and Optimization of Processes in Fluid Mechanics," September 4–7, 2004, Abrau-Dyurso, Russia, pp. 31–33 [in Russian].
[110] Gurchenkov A. A., Gridina E. D., and Bashlykov A. M. The problem of oscillations of a rotor containing a fluid with free surface // Zh. Vych. Mat. Mat. Fiz., 2001, vol. 44, no. 10. pp. 593–602 [in Russian].
[111] Gurchenkov A. A., Gridina E. D., and Bashlykov A. M. The problem of oscillations of a rotor containing a fluid with free surface // Zh. Vych. Mat. Mat. Fiz., 2002, vol. 42. no. 1. pp. 101–105 [in Russian].
[112] Gurchenkov A. A., Gridina E. D., and Bashlykov A. M. Mathematical Modeling of Filtration Processes in Rotor Systems. Moscow: Mashinostroenie Informatsionnye Tekhnologii, 2001 [in Russian].
[113] Gurchenkov A. A., Eleonskii V. M., and Kulagin N. E. Layer Structures in Nonlinear Vector Fields. Moscow: Vychisl. Tsentr, Ross. Akad. Nauk, 2007 [in Russian].
[114] Gurchenkov A. A., Esenkov A. S., and Tsurkov V. I. Control of the motion of a rotor with a cavity containing an ideal fluid, 1 // Izv. Ross. Akad. Nauk. Teor. Sist. Upr., 2006, no. 1. pp. 135–142 [in Russian].
[115] Gurchenkov A. A., Esenkov A. S., and Tsurkov V. I. Motion control of a rotor with a cavity with a viscous fluid // Avtom. Telemekh., 2007, no. 2, pp. 81–94 [Autom. Remote Control 2007, vol. 68, no. 2, pp. 284–295].
[116] Gurchenkov A. A., Esenkov A. S., and Tsurkov V. I. Control of the motion of a rotor with a cavity containing an ideal fluid, 2 // Izv. Ross. Akad. Nauk. Teor. Sist. Upr., 2006, no. 3. pp. 82–89 [in Russian].
[117] Gurchenkov A. A., Ivanov I. M., and Kuzovlev D. I. Stability of rotating bodies with fluid-containing cavities // Fundamental Problems of System Security, Moscow: Vuzovskaya Kniga, 2008, pp. 515–526 [in Russian].
[118] Gurchenkov A. A., Ivanov I. M., Kuzovlev D. I., and Nosov M. V. Analysis of stability problems and control of gyroscopic stabilization systems in spacecrafts // XXXIV International youth conference "Gagarin Readings," 2008, vol. 5, p. 68 [in Russian].
[119] Gurchenkov A. A., Ivanov I. M., Kuzovlev D. I., and Nosov M. V. Weakly perturbed motion of a fluid-filled top and a control problem // XVII All-Russia conference "Theoretical Foundations and Construction of Numerical Algorithms and Solution of Problems of Mathematical Physics with Application to Multiprocessor Systems" dedicated to the memory of K. I. Babenko, Abrau–Dyurso, September 15–21, 2008, pp. 143–144 [in Russian].
[120] Gurchenkov A. A., Ivanov I. M., Kulagin N. E., and Nosov M. V. The Problem of Stability and Control of the Motion of a Top Partially Filled with a Fluid, Moscow: Vychisl. Tsentr, Ross. Akad. Nauk, 2008 [in Russian].
[121] Gurchenkov A. A., Ivanov I. M., and Nosov M. V. Librational motion of a body in a viscous fluid // Dynamics of Inhomogeneous Systems, Tr. Inst. Sist. Anal. Ross. Akad. Nauk, 2010, vol. 49, no. 1, pp. 92–98 [in Russian].

[122] Gurchenkov A. A. and Korneev V. V. Control problem for a rotor containing a viscous fluid // Dynamics of Linear and Nonlinear Systems, Tr. Inst. Sist. Anal. Ross. Akad. Nauk, 2006, vol. 25, no. 2, pp. 10–26 [in Russian].

[123] Gurchenkov A. A., Korneev V. V., and Nosov M. V. Dynamics of weakly perturbed motion of a gyroscope filled with liquid and control problem // Prikl. Mat. Mekh., 2008, vol. 72. no. 6. pp. 904–911 [J. Appl. Math. Mech. 2008, vol. 72, no. 6, pp. 653–659].

[124] Gurchenkov A. A., Korneev V. V., and Nosov M. V. Control of the motion of a fluid-filled top // Dynamics of Inhomogeneous Systems, Tr. Inst. Sist. Anal. Ross. Akad. Nauk, 2007, vol. 10, no. 2, pp. 27–34 [in Russian].

[125] Gurchenkov A. A., Korneev V. V., and Nosov M. V. Stability and Control of Motion of a Fluid-Containing Top. Moscow: Vychisl. Tsentr, Ross. Akad. Nauk, 2006 [in Russian].

[126] Gurchenkov A. A., Kostikov A. A., Latyshev A. V., and Yushkanov A. A. Velocity of quantum Fermi-gas in Kramers' problem with accomodation of boundary conditions // Dynamics Inhomogeneous Systems, Tr. Inst. Sist. Anal. Ross. Akad. Nauk, 2008, vol. 32, no. 1, pp. 45–53 [in Russian].

[127] Gurchenkov A. A. and Kulagin N. E. Analytical Study of Gas-Kinetic and Hydrodynamic Problems, Moscow: Mosk. Gos. Oblast. Univ., 2003 [in Russian].

[128] Gurchenkov A. A. and Kulagin N. E. Localized and Periodic Solutions in Models of Nonlinear Scalar Field, Moscow: Vychisl. Tsentr, Ross. Akad. Nauk, 2004 [in Russian].

[129] Gurchenkov A. A. and Kulagin N. E. Symmetry Patterns in Simple Scalar Field Models, Moscow: Vychisl. Tsentr, Ross. Akad. Nauk. 2005 [in Russian].

[130] Gurchenkov A. A. and Latyshev A. V. Equations of the rotational motion of a solid body with a cavity containing a viscous fluid and inertial coupling coefficients // Physical Kinetics and Hydromechanics of Dispersion Systems, Available from VINITI, no. 5321-V86, Moscow, 1986 [in Russian].

[131] Gurchenkov A. A. and Nosov M. V. The problem of controlling the rotational motion of a fluid-filled rigid body // XXXII International youth conference "Gagarin Readings," Moscow, 2006, pp. 142–143 [in Russian].

[132] Gurchenkov A. A. and Nosov M. V. Optimal control of the motion of a fluid-filled top // IV All-Russia scientific conference of young scientists and students "State-of-the-Art and Priorities in the Development of Fundamental Sciences in Regions," Anapa, 2007, vol. 2, pp. 125–127 [in Russian].

[133] Gurchenkov A. A. and Nosov M. V. Stability of a Rotor with a Viscous Fluid, Moscow: Vychisl. Tsentr, Ross. Akad. Nauk, 2005 [in Russian].

[134] Gurchenkov A. A. and Yalamov Yu. I. Energy Dissipation in a Vibrating Structurally Heterogeneous Cavity Filled with a Viscous Liquid // Dokl. Akad. Nauk, 2002, vol. 382. no. 4. pp. 470–473 [Dokl. Phys., 2002, vol. 47, no. 2, pp. 99–101].

[135] Gurchenkov A. A. and Yalamov Yu. I. Unsteady Viscous Fluid Flow between Rotating Parallel Walls with Allowance for Thermal Slip along One of Them // Doklady Ross. Akad. Nauk, 2002, vol. 382. no. 1. pp. 54–57 [Dokl. Phys., 2002, vol. 47, no. 1, pp. 25–28].

[136] Gurchenkov A. A. Analytic solution of the model vector kinetic equations and their applications. 13th International conference on transport theory. Riccione (Italy), May 10–14, 1993.

[137] Gurchenkov A. A. The experimental analysis of bracking characters for the machines of the rotating type. Accepted for presentation at Inauqural Workshop of the International Institute for General Systems Studies. July 13–15, 1995. USA.

[138] Hall M. G. Vortex breakdown // Ann. Rev. Fluid Mech. 1972. vol. 4. pp. 195–218.

[139] Helmholtz H. L. F. // Variational Principles of mechanics, Collection of Papers, Moscow: Fizmatgiz, 1959 [in Russian].

[140] Hilbert D. and Courant R. Methods of Mathematical Physics, vol. II, New York: Interscience, 1962.

[141] Hough. S. S. The Oscillations of a Rotating Ellipsoidal Shell containing Fluid. Phil. Transactions (A), vol. 186, 1, 1895.

[142] Hung R. J., Lee C. C., and Leslie F. W. // Acta astronaut, no. 3, 1944, U.K. pp. 199–209.

[143] Hutton R. E. An Investigation of Nonlinear, Nonplaner Oscillations of Fluid in a Cylindrical Container. AIAA Fifth Annual Structures and Materials Conference, 1964, pp. 184–190.

[144] Ievleva O. B. Small oscillations of a pendulum having a spherical cavity filled with a viscous fluid // Prikl. Mat. Mekh., vol. 28, no. 6, pp. 1132–1134, 1964 [PMM, J. Appl. Math. Mech. vol. 28, pp. 1359–1361, 1964].

[145] Ievleva O. B. On oscillations of a body filled with a viscous fluid // Prikl. Mekh. Techn. Fiz., no. 6, 1966 [in Russian].

[146] Il'inskii V. S. Protection of Apparatuses fagainst Dynamic Actions. Moscow: Energiya, 1971 [in Russian].

[147] Ishlinskii A. Yu. Mechanics of Gyroscopic Systems. Moscow: Izd. Akad. Nauk SSSR, 1963 [in Russian].

[148] Ishlinskii A. Yu. and Temchenko M. E. On small oscillations of the vertical axis of a top with a cavity completely filled with an ideal incompressible fluid // Prikl. Mekh. Techn. Fiz., no. 3, 1960 [in Russian].

[149] Ishmukhametov A. Z. A dual regularized method for solving a class of convex minimization problems // Zh. Vychisl. Mat. Mat. Fiz. 2000. vol. 40, no. 7. pp. 1045–1060 [Comput. Math. Math. Phys. 2000, vol. 40, no. 7, pp. 1001–1016].

[150] Ishmukhametov A. Z. Regularized optimization methods with finite-step interior algorithms // Dokl. Ross. Akad. Nauk, 2003. vol. 390, no. 3, pp. 304–308 [Dokl. Math. 2003, vol. 67, no. 3, pp. 352–356].

[151] Ivanov M. I. Natural harmonic oscillations of a heavy fluid in basins of complex shape // Mekh. Zhidk. Gaza 2006, no. 1, pp. 131–148 [Fluid Dyn. 2006, vol. 41, no. 1, pp. 121–136].

[152] Ivchenko I. N. and Yalamov Yu. I. The kinetic theory of gas flow above a rigid wall in a velocity gradient field // Izv. Akad. Nauk SSSR, Mekh. Zhidk. Gaza, 1968, no. 6, pp. 139–143 [in Russian].

[153] Ivchenko I. N. and Yalamov Yu. I. Heat sliding of an inhomogeneous heated gas along a rigid plane surface // Izv. Akad. Nauk SSSR, Mekh. Zhidk. Gaza, 1969, no. 6, pp. 59–66 [in Russian].

[154] Jeffreys H. The Three Oscillations of Water in an Elliptical Lake. Proc. London Math. Soc., 1964, vol. 24.

[155] Kalitkin N. N. Numerical Methods, Moscow: Nauka, 1978 [in Russian].

[156] Kalman R. E. On the general theory of control systems // Proc. 1 IFAC Cong. (Moscow 1960). London: Butherworths, 1960. pp. 481–492.

[157] Kamke E. Gewöhnliche Differentialgleichungen, Leipzig: Acad. Verlag, 1959.

[158] Kapitonov B. V. Potential theory for the equation of small oscillations of a rotating fluid // Mat. Sb., 1979, vol. 109, no. 4. pp. 607–628 [in Russian].

[159] Karaketyan A. V. and Lagutina I. S. On the stability of uniform rotations of a top suspended on a string with account of dissipative constant moments // Izv. Ross. Akad. Nauk. Mekh. Tverd. Tela, 2000. no. 1. pp. 53–68 [in Russian].

[160] Karapetyan A. V. and Prokonina O. V. The stability of permanent rotations of a top with cavity filled with liquid on the plane with friction // Izv. Ross. Akad. Nauk. Prikl. Mat. Mekh., 2000. no. 1. pp. 85–94 [J. Appl. Math. Mech. 2000, vol. 64, no. 1, pp. 81–86].

[161] Kareva I. E. and Sennitskii V. L. Motion of a circular cylinder in a vibrating liquid // Prikl. Mekh. Techn. Fiz., 2001, vol. 42, no. 2. pp. 103–105 [J. Appl. Mech. Tech. Phys. 2001, vol. 42, no. 2, pp. 276–278].

[162] Karyakin V. E., Karyakin Yu. E., and Nesterov A. Ya. Calculation of circular flows of a viscous fluid in axially symmetric channels of any shape // Vesti Akad. Nauk SSSR, Fiz. Energ. N. 1990, no. 2, C. 82–88 [in Russian].
[163] Kelvin, Lord. Mathematical and Physical Papers, vol. IV. Cambridge, 1882.
[164] Klots Cornelius E. Evaporation from small particles // J. Phys. Chem., vol. 92, no. 21, 1988, pp. 5864–5868.
[165] Kochin N. E., Kubel' I. A., and Poze N. V. Theoretical Hydromechanics, vols. I, II, Moscow: Fizmatgiz, 1963 [in Russian].
[166] Kogan V. B. Theoretical Foundations of Typical Processes in Chemical Technology. Moscow: Khimiya, 1977 [in Russian].
[167] Kokotovic P. and Arcak M. Constructive nonlinear control: a historical perspective // Automatica. vol. 37. no. 5. pp. 637–662.
[168] Kolesnikov N. N. On the stability of a free rigid body with a cavity filled with an incompressible viscous fluid // Prikl. Mat. Mekh., vol. 26, no. 4, pp. 774–777, 1962 [PMM, J. Appl. Math. Mech., 1962, vol. 26, pp. 914–923].
[169] Kopachevskii N. D. Small Motions and Free Oscillations of an Ideal Rotating Fluid, Preprint no. 38–71, Institute for Low Temperature Physics and Engineering, Acad. Sci. Ukr. SSR, Kharkov, 1978 [in Russian].
[170] Kopachevskii N. D. and Myshkis A. D. Hydrodynamics in weak force fields. On small oscillations of a viscous fluid in the potential field of mass forces // Zh. Vychisl. Mat. Mat. Fiz., vol. 8, no. 6, 1966 [in Russian].
[171] Korn G. A. and Korn T. M., Mathematical Handbook for Scientists and Engineers, New York: McGraw-Hill, 1961.
[172] Koterov V. N. and Shmyglevskii Yu. D. Stokes plane-parallel vortex systems in channels // Izv. Ross. Akad. Nauk. Mekh. Zhidk. Gaz., 2000, no. 5. pp. 57–66 [Fluid Dyn. 2000, vol. 35, no. 5, pp. 674–681].
[173] Kovalev A. M. The controllability of dynamical systems with respect to part of the variables // Prikl. Mat. Mekh., 1993, vol. 57. no. 6. pp. 41–50 [J. Appl. Math. Mech. 1993, vol. 57, no. 6, pp. 995–1004].
[174] Kovalev A. M. Partial stability and stabilization of dynamical systems // Ukr. Mat. Zh. 1995, vol. 47. no. 2. pp. 186–193 [Ukr. Math. J. 1995, vol. 47, no. 2, pp. 218–226].
[175] Kovalyev A. M. Control and stabilization problems with respect to a part of the variables // ZAMM, 1994, vol. 74. no. 7. pp. 59–60.
[176] Krapivina E. N. and Lyubimova T. P. Nonlinear regimes of convection of a viscoelastic fluid in a closed cavity heated from below // Izv. Ross. Akad. Nauk. Mekh. Zhidk. Gaz., 2000. no. 4. pp. 5–18 [Fluid Dyn. 2000, vol. 35, no. 4, pp. 473–478].
[177] Krasnoshchekov P. S. Small oscillations of a solid body having cavities filled with viscous fluid // Numerical Methods for Solving Problems of Mathematical Physics, Moscow: Nauka, 1966 [in Russian].
[178] Krasnoshchekov P. S. On oscillations of a physical pendulum having a cavity filled with a viscous fluid. Prikl. Mat. Mekh., vol. 27, no. 2, 1963 [in Russian].
[179] Krasovskii N. N. Game Problems of Meeting of Motions. Moscow: Nauka, 1970 [in Russian].
[180] Krasovskii N. N. To optimal control theory // Avtom. Telemekh., 1957. vol. 18. no. 11. pp. 960–970 [in Russian].
[181] Krasovskii N. N. Some problem theory stability motion. Moscow: Fizmatlit, 1959.
[182] Krasovskii N. N. Stabilization problems for controlled motion // Appendix 4 in: Malkin I. G. Motion Stability Theory. Moscow: Naukascience, 1966. pp. 475–514 [in Russian].
[183] Krasovskii N. N. Motion Control Theory. Moscow: Nauka, 1968.
[184] Krasovskiy N. N and Subbotin A. I. Game-Theoretical Control Problems. New York: Springer–Verlag, 1987.

[185] Krein S. G. Differential equations in Banach spaces and their applications to hydromechanics. Usp. Mat. Nauk, vol. 12, no. 1, 1957 [in Russian].
[186] Krein S. G. On oscillations of a viscous fluid in a vessel // Dokl. Akad. Nauk SSSR, vol. 159, no. 2, 1964 [in Russian].
[187] Krein S. G. and Moiseev N. N. On oscillations of a solid body containing a fluid with free surface // Prikl. Mat. Mekh., vol. 21, no. 2, 1957 [in Russian].
[188] Krivtsov A. M. The influence of a bounded rotational moment on the stability of the steady motion of an asymmetric top // Izv. Ross. Akad. Nauk. Mekh. Tverd. Tela, 2000. no. 2. pp. 33–44 [in Russian].
[189] Krushinskaya S. I. Oscillations of a heavy viscous fluid in a moving vessel // Zh. Vychisl. Mat. Mat. Fiz. vol. 5, no. 3, pp. 519–536 (1965) [U.S.S.R. Comput. Math. Math. Phys. 5(1965), No. 3, 168–192 (1967)].
[190] Krylov A. N. Ship Roll: Collection of Works, vol. XI, Izd. Akad. Nauk SSSR, 1951 [in Russian].
[191] Kuzin A. K. Quasi-cylindrical figures of equilibrium of a rotating fluid // Izv. Ross. Akad. Nauk. Mekh. Zhidk. Gaza, 2000. no. 3. pp. 22–38 [Fluid Dyn. vol. 35, No. 3, pp. 331–338 (2000)].
[192] Kurzhanskii A. B. Control and Observation under Uncertainty Conditions. Moscow: Nauka, 1977 [in Russian].
[193] Kurosh A. G. A Course in Higher Algebra. Moscow: Nauka, 1968 [in Russian].
[194] Ladyzhenskaya O. A. Boundary Value Problems of Mathematical Physics. Moscow: Nauka, 1973 [in Russian].
[195] Lakenath Debnath, Unsteady multiple boundary Sukla Mukherjee layers on a porous plate in a rotating system. Journal Phys. Fluids, vol. 16, p. 9, 1973.
[196] Lakshmikantham V. and Matrosov V. M., Sivasundaram S. Vector Lyapunov Function and Stability Analysis of Nonlinear Systems. Dordrecht: Kluwer Acad. Publ., 1991.
[197] Lamb H. Hydrodynamics. Cambridge: Cambridge Univ. Press, 1932.
[198] Landau L. D. and Lifshits E. M. Continuum Mechanics, Moscow: Gostekhizdat, 1955 [in Russian].
[199] Lavrent'ev M. A. and Shabat V. B. Methods of the Theory of Functions of a Complex Variable. Moscow: Fizmatgiz, 1958 [in Russian].
[200] Lawrence H. R., Wang C. I., and Reddy R. B. Variational Solution of Fuel Sloshing Modes. Jet Prop., vol. 28, no. 11, 1958.
[201] Lee C. C. and Leslie F. W. Dynamic characteristics of the partially filled rotating dewar of the gravity probe–B spacecraft // Acta astronaut. 1944. vol. 32. no. 3. pp. 199–209.
[202] Leibovich S. The structure of vortex breakdown // Ann. Rev. Fluid Mech. 1978. vol. 10. pp. 221–246.
[203] Leibovich S. Vortex stability and breakdown: survey and extension // AIAA Journal. 1984. vol. 22. no. 9. pp. 1192–1206.
[204] Levin E. Oscillations of Fluid in a Rectilinear Conical Container. AIAA Journal, 1963, vol. 1, no. 6.
[205] Lighthill J. Waves in Fluids. Cambridge: Cambridge Univ. Press, 1978.
[206] Loitsyanskii L. V. and Lur'e A. I. A Course in Theoretical Mechanics. Gostekhteorizdat, 1940, vol. II [in Russian].
[207] Lur'e A. I. Nonlinear Problems of Automatic Control Theory. Moscow: Gostekhizdat, 1951 [in Russian].
[208] Lyapunov A. M. General Motion Stability Problem. Moscow: GITTL, 1950 [in Russian].
[209] Makarenko V. G. and Tarasov V. F. On the flow structure of a rotating fluid after a motion of a body in it // Prikl. Mekh. Techn. Fiz. 1988, no. 6, pp. 113–117 [in Russian].
[210] Malashenko S. V. and Temchenko M. E. A method for experimentally studying the stability of the motion of a top with a fluid-filled cavity inside // Prikl. Mekh. Techn. Fiz., no. 3, 1960 [in Russian].

[211] Malkin I. G. Theory of Motion Stability. Moscow: Gostekhizdat, 1952 [in Russian].
[212] Maslennikova V. N. Mathematical Questions of the Hydrodynamics of a Rotating Fluid and a Sobolev System, Summary of doctoral dissertation in physics and mathematics, Novosibirsk, 1971 [in Russian].
[213] Matveev Yu. L. and Matveev L. T. Special features of the formation, development, and motion of tropical cyclones // Izv. Ross. Akad. Nauk. Fiz. Atm. Okeana, 2000, no. 6, pp. 760–767 [in Russian].
[214] Matrosov V. M. Method of Lyapunov Vector Functions: Analysis of Dynamical Properties of Nonlinear Systems. Moscow: Nauka, 2001 [in Russian].
[215] Matrosov V. M., Anapol'skii L. Yu., and Vasil'ev S. N. Comparison Method in Mathematical System Theory. Novosibirsk: Nauka, 1980 [in Russian].
[216] Mauwell A. R. A variational principle for steady nomenergetic compressible flow with finite shocks // Wave motion. 1980. vol. 2. no. 1. pp. 83–85.
[217] Mikhailov D. N. and Nikolaevskii V. N. Dynamics of flow through porous media with unsteady phase permeabilities // Izv. Ross. Akad. Nauk. Mekh. Zhidk. Gaza, 2000. no. 5. pp. 103–127 [Fluid Dyn. 2000, vol. 35, no. 5, pp. 715–724].
[218] Mikhlin S. G. Variational Methods in Mathematical Physics. Moscow: GITTL, 1957 [in Russian].
[219] Mikishev G. N. and Dorozhkin N. Ya. Experimental Study of Free Oscillations of a Fluid in Vessels // Izv Akad. Nauk SSSR. Mekh. Mashinostr., no. 4, 1961 [in Russian].
[220] Mikishev G. N., Nevskaya E. A., Mel'nikov I. M., and Dorozhkin N. Ya. Experimental study of the perturbed motion of a rigid body with cavities partially filled with fluid // Kosmich. Issled. vol. 3, no. 2, 1965 [in Russian].
[221] Mikishev G. N. and Rabinovich B. I. Dynamics of a Rigid Body with Cavities Partially Filled with a Fluid. Moscow: Mashinostroenie, 1968 [in Russian].
[222] Miles J. W. Free-surface oscillations in a slowly rotating liquid. Journal fluid mech., vol. 18, no. 2, 1964.
[223] Miles J. W. Ring Damping of Free Surface Oscillations in a Circulal Tank. Journal of Appl. Mech., 1958, vol. 25, no. 6.
[224] Miroshnik I. V., Nikiforov V. O., and Fradkov A. L. Nonlinear and Adaptive Control of Complex Dynamic Systems. St. Petersburg: Nauka, 2000.
[225] Moiseev N. N. Motion of a Solid Body with a Cavity Partially Filled with an Ideal Dropping Liquid // Dokl. Akad. Nauk SSSR, vol. 85, no. 4, 1952 [in Russian].
[226] Moiseev N. N. Problem about the Motion of a Solid Body Containing Liquid Masses with Free Surface // Mat. Sbornik, vol. 32, no. 1, pp. 61–96, 1953 [in Russian].
[227] Moiseev N. N. Problem about Small Oscillations of an Open Fluid-Containing Vessel under the Action of an Elastic Force // Ukr. Mat. Zh., vol. 4, no. 2, pp. 168–173, 1952 [in Russian].
[228] Moiseev N. N. On oscillations of a heavy ideal incompressible fluid in a vessel // Dokl. Akad. Nauk SSSR, vol. 85, no. 5, pp. 719–722, 1952 [in Russian].
[229] Moiseev N. N. Boundary problems for linearized Navier-Stokes equations when the viscosity is small // Zh. Vychisl. Mat. Mat. Fiz., vol. 1, no. 3, pp. 548–550, 1961 [U.S.S.R. Comput. Math. Math. Phys. vol. 1, pp. 628–632, 1962].
[230] Moiseev N. N. Mathematical methods for studying nonlinear fluid oscillations // Proceedings of International Symposium on Nonlinear Oscillations. Kiev: Izd. Akad. Nauk Ukr. SSR, vol. 3, 1963 [in Russian].
[231] Moiseev N. N. and Petrov A. A. Numerical Methods for Calculating Eigenfrequencies of Oscillations of a Bounded Volume of Fluid. Moscow: Vychisl. Tsentr, Akad. Nauk SSSR, 1966 [in Russian].
[232] Moiseev N. N. and Rumyantsev V. V. Dynamics of a Body with Fluid-Containing Cavities, Moscow: Nauka, 1965 [in Russian].

[233] Moiseev N. N. and Chernous'ko F. L. Problems on the oscillation of a fluid subject to surface tension forces // Zh. Vychisl. Mat. Mat. Fiz. vol. 5, no. 6, pp. 1071–1095, 1965 [U.S.S.R. Comput. Math. Math. Phys. vol. 5, no. 6, pp. 128–160, 1965].

[234] Moisseyev N. N. Sur certains problemes mathematiques du mouvement relatif des satellites. Dynamics of satellites. Syposium, Paris, 1962. Springer-Verlag, Berlin–Gottingen–Heidelberg, 1963.

[235] Monin A. S. and Shishkov Yu. A. Circulation mechanisms of atmosphere climate fluctuations // Izv. Ross. Akad. Nauk. Fiz. Atm. Okeana, 2000. no. 1. pp. 27–34 [in Russian].

[236] Morse P. M. and Feshbach H. Methods of Theoretical Physics. New York: McGraw-Hill, 1953.

[237] Mory M. and Yurchenko N. Vortex generation by suction in a rotating tank // Eur. J. Mech. 1993. no. 6. pp. 729–747.

[238] Naimark M. A. Linear Differential Operators. New York: Ungar, 1967.

[239] Narimanov G. S. On the motion of a vessel partially filled with a fluid and allowing for the nonsmallness of the fluid motion // Prikl. Mat. Mekh., vol. 21, no. 4, 1957 [in Russian].

[240] Narimanov G. S. On the motion of a rigid body with a cavity partially filled with a fluid // Prikl. Mat. Mekh., vol. 20, no. 1, 1956 [in Russian].

[241] Neumann C. Gydrodynamishe Untersuchungen. Leipzig, 1883.

[242] O'Brien V. Steady spheroidal vortieces—more exact Solutions to the Navier–Stokes equation // Q. Appl. Math. 1961. vol. 19. no. 2. pp. 163–168.

[243] Okhotsimskii D. E. To the theory of rocket motion // Prikl. Mat. Mekh., 1940. vol. 10. no. 2 [in Russian].

[244] Okhotsimskii D. E. To the theory of motion of a body with cavities partially filled with a fluid // Prikl. Mat. Mekh., vol. 20, no. 1, 1956 [in Russian].

[245] Onishi Yoshimoto. On the behavior of a noncondensable gas of a small amount in weakly nonlinear evaporation and condensation of a vapor // J. Phys. Soc. Japan, vol. 55, no. 9, 1986, pp. 3080–3092.

[246] Oziraner A. S. On optimal stabilization of motion with respect to a part of variables // Prikl. Mat. Mekh., 1978, vol. 42. no. 2. pp. 272–276 [J. Appl. Math. Mech. vol. 42, pp. 284–289, 1979].

[247] Paley R. E. A. C. and Wiener N., Fourier Transforms in the Complex Domain, New York: Amer. Math. Soc., 1934.

[248] Petrov A. A. Approximate solution of problems about oscillations of a fluid in a cylindrical vessel with horizontal generatrix // Variational Methods in Problems about Oscillations of Fluids and Fluid-Containing Bodies, Moscow: Vychisl. Tsentr, Akad. Nauk SSSR, 1962. pp. 213–220 [in Russian].

[249] Petrov A. A. An approximate method for the calculation of characteristic oscillations of a liquid in vessels of arbitrary shape and the Zhukovskii potentials for these vessels // Zh. Vychisl. Mat. Mat. Fiz. 1963, vol. 3. no. 5. pp. 958–964 [U.S.S.R. Comput. Math. Math. Phys. vol. 3(1963), pp. 1307–1316, 1966].

[250] Petrov A. A., Popov Yu. P., and Pukhnachev Yu. V. Evaluation of the proper vibrations of a fluid in a fixed vessel by the variational method // Zh. Vychisl. Mat. Mat. Fiz. 1964, vol. 4. no. 5. pp. 880–895 [U.S.S.R. Comput. Math. Math. Phys. vol. 4 (1964), no. 5, pp. 123–144, 1967].

[251] Petrov A. G. The development of the flow of viscous and viscoplastic media between two parallel plates // Prikl. Mat. Mekh., 2000, vol. 64, issue. 1. pp. 127–136 [J. Appl. Math. Mech. vol. 64, no. 1, pp. 123–132, 2000].

[252] Petrov V. M. and Chernous'ko F. L. Determination of the equilibrium shape of a fluid under the action of gravity and surface tension forces // Izv. Akad. Nauk SSSR, Mekh. Zhidk. Gaz., no. 5, 1966 [in Russian].

[253] Poincarè H. Sur la prècèssion des corps dèformables. Bulletin astronomoque, t. XXVII, 1910.

[254] Pontryagin L. S., Boltyanskii V. G., Gamkrelidze R. V., and Mishchenko E. F. Mathematical Theory of Optimal Processes. Moscow: Fizmatlit. 1961 [in Russian].

[255] Pozharitzkii G. K. and Rumyantsev V. V. The problem of the minimum in the question of stability of motion of a solid body with a liquid-filled cavity // Prikl. Mat. Mekh., vol. 27, no. 1, pp. 16–26, 1963 [PMM, J. Appl. Math. Mech. vol. 27, pp. 18–32, 1963].

[256] Pozharitzkii G. K. and Rumyantsev V. V. On the influence of viscosity on the stability of equilibrium and steady rotations of a rigid body with a cavity partially filled with a viscous fluid // Prikl. Mat. Mekh., vol. 28, no. 1, 1964 [in Russian].

[257] Prudnikov A. P., Brychkov Yu. A., and Marichev O. I. Integrals and Series. Moscow: Nauka, 1981 [in Russian].

[258] Pyatnitskii E. S. Synthesis of stabilization systems of program motion for nonliner objects of control systems // Avtom. Telemekh., 1993, vol. 54. no. 7. pp. 19–37 [Autom. Remote Control vol. 54, no. 7, pt. 1, pp. 1046–1062, 1993].

[259] Rabinovich B. I. On equations of perturbed motion of a solid body with a cylindrical cavity partially filled with a fluid // Prikl. Mat. Mekh., vol. 20, no. 1, 1956 [in Russian].

[260] Rabinovich B. I., Dokuchaev L. V., and Polyakova Z. M. Calculation of coefficients in equations of perturbed motion of a solid body with cavities partially filled with a fluid // Kosmich. Issled. vol. 3, no. 2, 1965 [in Russian].

[261] Rabinovich B. I. and Rogovoi V. M. // Kosmich. Issled. 1970, vol. 8, no. 3, p. 315 [in Russian].

[262] Reed M. C. and Simon B. Methods of Modern Mathematical Physics, vol. 2. Academic Press, New York, 1975.

[263] Reist P. C. Introduction to Aerosol Science, New York: Macmillan, 1984.

[264] Riley J. D. Sloshing of Liquid in Spherical Tanks. Journal of the Aero/Space Sciences. 1961, vol. 28, no. 3.

[265] Rumyantsev V. V. Lyapunov methods in the study of motion stability of rigid bodies with fluid-filled cavities // Izv. Akad. Nauk SSSR, Mekh. Mashinostr, no. 6, 1963 [in Russian].

[266] Rumyantsev V. V. On the motion of a solid body containing a cavity filled with a viscous fluid // Prikl. Mat. Mekh., vol. 28, no. 6, 1964 [in Russian].

[267] Rumyantsev V. V. The optimality of motion stabilization in some of the variables // Izv. Ross. Akad. Nauk. Tekh. Kibern. no. 1. pp. 184–189, 1993 [J. Comput. Syst. Sci. Int. vol. 32, no. 2, pp. 151–156, 1994].

[268] Rumyantsev V. V. On the optimal stabilization of controlled systems // Prikl. Mat. Mekh., 1970. vol. 34. no. 3. pp. 440–456 [J. Appl. Math. Mech. vol. 34, pp. 415–430, 1970].

[269] Rumyantsev V. V. On equations of motion of a solid body with a fluid-filled cavity // Prikl. Mat. Mekh., vol. 19, no. 1, 1955 [in Russian].

[270] Rumyantsev V. V. The stability of the rotational motions of a solid body with a liquid cavity // Prikl. Mat. Mekh., vol. 23, pp. 1057–1065, 1959 [PMM, J. Appl. Math. Mech. vol. 23, pp. 1512–1524, 1960].

[271] Rumyantsev V. V. On the stability of rotation of a top with a cavity filled with a viscous fluid // Prikl. Mat. Mekh., vol. 24, no. 4, 1960 [in Russian].

[272] Rumyantsev V. V. On the stability of motion of a rigid body containing a fluid possessing surface tension // Prikl. Mat. Mekh., vol. 28, no. 4, pp. 746–753, 1964 [PMM, J. Appl. Math. Mech. vol. 28, pp. 908–916, 1964].

[273] Rumyantsev V. V. On stability of equilibria of a rigid body with liquid-filled cavities // Dokl. Akad. Nauk SSSR, vol. 124, no. 2, pp. 291–294, 1959 [Sov. Phys., Dokl. vol. 4, pp. 46–49, 1959].

[274] Rumyantsev V. V. Equations of motion of a solid body having a cavity incompletely filled with a fluid // Prikl. Mat. Mekh., vol. 18, no. 6, 1954 [in Russian].

[275] Rumyantsev V. V. Stability of rotation of a solid body with an ellipsoidal cavity filled with a fluid // Prikl. Mat. Mekh., vol. 21, no. 6, 1957 [in Russian].
[276] Rumyantsev V. V. On the stability with respect to a part of the variables // Symp. Math. Vol. 6. Meccanica non-lineare. Stability. 23–26 febbrario, 1970. New York: Acad. Press. 1971. pp. 243–265.
[277] Rvalov R. V. and Rogovoi V. M. On rotational motions of a body with a fluid-containing cavity // Izv. Akad. Nauk SSSR, Mekh. Tverd. Tela, 1972, no. 3, pp. 15–20 [in Russian].
[278] Sanochkin Yu. V. Viscosity effect on free surface waves in fluids // Izv. Ross. Akad. Nauk. Mekh. Zhidk. Gaza, 2000. no. 4. pp. 156–164 [Fluid Dyn. vol. 35, no. 4, pp. 599–604, 2000].
[279] Semenov E. V. Convergent laminar fluid flow between two rotating disks // Prikl. Mekh. Techn. Fiz., 2000. no. 2. pp. 77–83 [Prikl. Mekh. Tekh. Fiz. vol. 41, no. 2, pp. 77–83, 2000].
[280] Shen C. The concentration-jump coefficient in a rarefied binary gas mixture // J. Fluid Mech. vol. 137, pp. 221–231, 1983.
[281] Shestakov A. A. Generalized Lyapunov Direct Method for Systems with Distributed Parameters. Moscow: Nauka, 1990 [in Russian].
[282] Schmidt A. G. Oscillations of a viscous fluid of finite depth caused by an initial displacement of its free surface // Zh. VYchil. Mat. Mat. Fiz., vol. 5, no. 2, 1965 [in Russian].
[283] Shmyglevskii Yu. D. An Analytic Study of Fluid Dynamics. Moscow: Editorial URSS, 1999 [in Russian].
[284] Shmyglevskii Yu. D. Two examples of flow around plates in the Stokes approximation // Zh. Vychisl. Mat. Mat. Fiz. vol. 35, no. 6, pp. 997–1000, 1995 // [Comput. Math. Math. Phys. vol. 35, no. 6, pp. 795–798, 1995].
[285] Shmyglevskii Yu. D. Inversion of a singularity of the Navier-Stokes equations // Zh. Vychisl. Mat. Mat. Fiz. 1988, vol. 28, no. 11, pp. 1748–1750 [U.S.S.R. Comput. Math. Math. Phys., 1988, vol. 28, no. 6, pp. 103–105].
[286] Shmyglevskii Yu. D. On "vortex breakdown" // Izv. Ross. Akad. Nauk. Mekh. Zhidk. Gaza. 1995, no. 3, pp. 167–169 [Fluid Dyn. 1995, vol. 30, no. 3, pp. 477–478].
[287] Shmyglevskii Yu. D. Vortical formations in an ideal and a viscous fluid // Zh. Vychisl. Mat. Mat. Fiz. 1994, vol. 34, no. 6, pp. 955–959 [Comput. Math. Math. Phys. 1994, vol. 34, no. 6, pp. 827–831].
[288] Shmyglevskii Yu. D. On vortical formations in plane-parallel flows of a viscous and an ideal fluid // Izv. Ross. Akad. Nauk. Mekh. Zhidk. Gaza. 1977, no. 6, pp. 88–92 [in Russian].
[289] Shmyglevskii Yu. D. Spiral flows of ideal and viscous liquids // Zh. Vychisl. Mat. Mat. Fiz. 1993, vol. 33, no. 12, pp. 1905–1911 [Comput. Math. Math. Phys. 1993, vol. 33, no. 12, pp.1665–1670].
[290] Shmyglevskii Yu. D. Viscous liquid flows that are independent of the Reynolds number // Zh. Vychisl. Mat. Mat. Fiz. 1990, vol. 30, no. 6, pp. 951–955 [U.S.S.R. Comput. Math. Math. Phys. 1990, vol. 30, no. 3, pp. 220–223].
[291] Shmyglevskii Yu. D. On chains of axially symmetric vortices // Izv. Ross. Akad. Nauk. Mekh. Zhidk. Gaza. 1997, no. 2, pp. 174–176 [in Russian].
[292] Shmyglevskii Yu. D. On isobaric planar flows of a viscous incompressible liquid // Zh. Vychisl. Mat. Mat. Fiz. 1985, vol. 25, no. 12, pp. 1895–1898 [U.S.S.R. Comput. Math. Math. Phys. 1985, vol. 25, no. 6, pp. 191–193].
[293] Shmyglevskii Yu. D. On an inertial flow // Zh. Vychisl. Mat. Mat. Fiz. 1990, vol. 30, no. 12, pp. 1833–1834 [U.S.S.R. Comput. Math. Math. Phys. 1990, vol. 30, no. 6, 167–168].

[294] Shmyglevskii Yu. D. and Shcheprov A. V. Axisymmetric vortex formations in a viscous liquid // Zh. Vychisl. Mat. Mat. Fiz. 1995, vol. 35, no. 3, pp. 472–476 [Comput. Math. Math. Phys. 1995, vol. 35, no. 3, pp. 379–382].

[295] Siljak D. Decentralized control of complex systems. Boston, Mass.: Academic Press, 1990.

[296] Slezkin N. A. Dynamics of Viscous Incompressible Fluid. Moscow: Gostekhizdat, 1955 [in Russian].

[297] Smirnov V. I. A Course in Higher Mathematics, vol. 2. GONTI, 1938 [in Russian].

[298] Sneddon I. N., Fourier Transforms, New York: McGraw-Hill, 1951.

[299] Sobolev S. L. On the motion of a symmetric top with a fluid-filled cavity // Prikl. Mat. Mekh. 1960. no. 3. pp. 20–55 [in Russian].

[300] Sobolev S. L. Equations of Mathematical Physics. GITTL, 1950 [in Russian].

[301] Soda T. A kinetic theory analysis of unsteady evaporation from a liquid surface with temperature change // J. Phys. Soc. Jap. vol. 55, no. 5, 1968, pp. 1556–1567.

[302] Spall R. E., Gatski T. B., and Ash R. L. The structure and dynamics of bubble type vortex breakdown // Proc. Roy. Soc. 1990. vol. 429. no. 1877. pp. 613–637.

[303] Stewartson K. On the Stability of a Spinning Top Containing Liquid. J. Fluid Mech., vol. 5, 4, 1959.

[304] Stewartson K. and Roberts P. H. On the motion of a fluid in a spheroidal cavity of a precessing rigid body. Journal of fluid mech., vol. 17, part 1, 1963.

[305] Stokes G. Mathematical and Physical Papers, vol. I. Cambridge, 1880.

[306] Sundaram P., Kurosat M., and Wu J. M. Vortex dynamics analysis of unsteady vortex waves // AIAA Journal. 1991. no. 3. pp. 321–326.

[307] Sychev V. V. Asymptotic theory of vortex breakdown // Izv. Ross. Akad. Nauk. Mekh. Zhidk. Gaza. 1993. no. 3. pp. 78–90 [Fluid Dyn. vol. 28, no. 3, pp. 356–364, 1993].

[308] Tkhai V. N. The stability of regular Grioli precessions // Izv. Ross. Akad. Nauk. Prikl. Mat. Mekh., 2000, vol. 64, no. 5, pp. 848–859 [J. Appl. Math. Mech. 2000, vol. 64, no. 5, pp. 811–819].

[309] Thornley C. On Stokes and Rayleigh layers in a rotating system. Journal Mech. Appl. Math., vol. XXI, p. 4, 1968.

[310] Tikhonov A. N. and Arsenin V. Ya. Methods for Solving Ill-Posed Problems. Moscow: Nauka, 1986.

[311] Trigub V. N. The problem of breakdown of a vortex line // Prikl. Mat. Mekh. 1985, vol. 49, no. 2, pp. 220–226 [J. Appl. Math. Mech. 1985, vol. 49, pp. 166–171].

[312] Tsurkov V. I. Dynamic Problems of Large Dimension. Moscow: Nauka, 1988 [in Russian].

[313] Van Dyke M. An album on fluid motion. Standford. Parabolic Press. 1982. p. 176.

[314] Van Dyke M. Perturbation methods in fluid mechanics. N.Y. London. Acad. Press. 1964. p. 229.

[315] Variational Methods in Problems on Oscillations of a Fluid and a Fluid-Containing Body, Collection of Papers. Moscow: Vychisl. Tsentr, Akas. Nauk SSSR, 1962 [in Russian].

[316] Vasil'ev S. N., Zherlov A. K., Fedosov E. A., and Fedunov B. E. Intelligent Control of Dynamic Systems. Moscow: Fizmatlit, 2000 [in Russian].

[317] Vasil'ev F. P. Optimization Methods. Moscow: Faktorial, 2002 [in Russian].

[318] Vishik M. I. and Lyusternik L. A. Regular degeneration and boundary layer for linear differential equations with a small parameter // Usp. Mat. Nauk, vol. 12, no. 5, pp. 2–122, 1957 [in Russian].

[319] Vorotnikov V. I. To the nonlinear game problem of reorientation of an asymmetric rigid body // Izv. Ross. Akad. Nauk. Mekh. Tverd. Tela, 1999, no. 1. pp. 3–18 [in Russian].

[320] Vorotnikov V. I. Stability of Dynamic Systems with Respect to a Part of Variables, Moscow: Nauka, 1991 [in Russian].

[321] Vorotnikov V. I. Partial Stability and Control. Boston: Birkhauser, 1998.

[322] Watson G. N. Treatise on the Theory of Bessel Functions. Cambridge: Cambridge Univ. Press, 1944.

[323] Wilner L. B., Morrison W. L., and Brown A. E. An Instrument for Measuring Liquid Level and Slosh in the Tanks of a Liquid–Propellant Rocket. Proc. of the IRE, 1960, vol. 48, no. 4.

[324] Yakimov Yu. L. The whirlwind and a singular limiting solution of the Navier-Stokes equations // Izv. Ross. Akad. Nauk. Mekh. Zhidk. Gaza. 1988, no. 6, pp. 28–33 [Fluid Dyn. 1988, vol. 23, no. 6, pp. 819–828].

[325] Yalamov Yu. I., Gaidukov M. N., and Yushkanov A. A. Isothermal sliding of a binary gas mixture along a plane surface // Inzh.-Fiz. Zh., 1975, vol. 29, no. 3, pp. 489–493 [in Russian].

[326] Yalamov Yu. I. and Sanasaryan A. S. Drop motion in a temperature-inhomogeneous viscous medium // Inzh.-Fiz. Zh., 1975, vol. 28, no. 6 [in Russian].

[327] Yalamov Yu. I., and Sanasaryan A. S. Diffusiphoresis of coarse and moderately coarse drops in viscous media // Zh. Tekhn. Fiz., 1977, vol. 47, no. 5, pp. 1063–1066 [in Russian].

[328] Yalamov Yu. I. and Sanasaryan A. S. Motion of coarse drops, rigid particles, and gas bubbles in temperature-inhomogeneous liquids and gases in the sliding mode // Zh. Tekhn. Fiz., 1975, vol. 47, no. 5, pp. 1063–1066 [in Russian].

[329] Yalamov Yu. I. and Yushkanov A. A. Heat sliding of a binary gas mixture on a curved surface // Physics of Aerodisperse Systems and Physical Kinetics, Moscow: Mosk. Oblast. Pedagog. Univ. im. N. K. Krupskoi, 1978, issue 2, pp. 162–175 [in Russian].

[330] Zaslavskii M. M. and Perfil'ev V. A. Hamilton's principle for a model of a nonviscous contractible fluid in Euler coordinates // Izv. Akad. Nauk SSSR. Mekh. Zhidk. Gaz. 1969. no. 1. pp. 105–109 [in Russian].

[331] Zharov V. A. Determination of slippage speed for a binary mixture of gases // Izv. Akad. Nauk SSSR, Mekh. Zhidk. Gaz., no. 2, 1972. pp. 98–104 [in Russian].

[332] Zhukovskii N. E. On the motion of a solid body having a cavity filled with a homogeneous dropping fluid // Collection of Works, vol. II, Moscow: Gostekhizdat, 1947 [in Russian].

[333] Zil'bergleit A. S. Exact solution of a nonlinear system of partial differential equations in hydrodynamics // Dokl. Ross. Akad. Nauk, 1993. vol. 328. no. 5. pp. 564–566 [Phys.-Dokl. 1993, vol. 38, no. 2, pp. 61–63].

[334] Zubov V. I. Mathematical Methods for Studying Automatic Control Systems. Leningrad: Sudpromgiz, 1959 [in Russian].

[335] Zubov V. I. A. M. Lyapunov's Methods and Their Application. Leningrad: Izd. Leningrad Gos. Univ., 1957 [in Russian].

Subject index

Communications in Cybernetics, Systems Science and Engineering

Book Series Editor: Jeffrey 'Yi-Lin' Forrest

ISSN: 2164-9693

Publisher: CRC Press / Balkema, Taylor & Francis Group

1. A Systemic Perspective on Cognition and Mathematics
   Jeffrey Yi-Lin Forrest
   ISBN: 978-1-138-00016-2 (Hb)

2. Control of Fluid-Containing Rotating Rigid Bodies
   Anatoly A. Gurchenkov, Mikhail V. Nosov & Vladimir I. Tsurkov
   ISBN: 978-1-138-00021-6 (Hb)